COLORADO MINING STORIES

Hazards, Heroics, and Humor

CAROLINE ARLEN

WESTERN REFLECTIONS PUBLISHING COMPANY

ISBN 1-890437-74-3

Cover and text design by Paulette Livers, Boulder

WESTERN REFLECTIONS PUBLISHING COMPANY
P.O. Box 1647
Montrose, CO 81401

www.westernreflectionspub.com
<http://www.westernreflectionspub.com>

CONTENTS

INTRODUCTION

The thing to remember, is that miners lie sometimes.
—Myron Jones

The concept for this book began in 1997 at the Hardrockers Holidays, an annual gathering of miners in Silverton, Colorado. The Hardrockers event is a sort of combination reunion and competition—who can drill the fastest and so on. Friends, relatives, and admirers sit on the overlooking hill, drinking beer and reminiscing.

I knew one of the miners. I sat with him and his friends and listened to them tell stories. I did a lot of listening that day. The stories ran like rivers, and I caught a glimpse of a whole colorful world I had only imagined as taking place in old black and white photographs. I decided to try to document their stories in their own words.

Hardrock miners in the San Juans tend to be very proud. Mining required a great deal of arduously achieved experience and skill. Compared to other occupations available to them, the job paid very well. Supposing a miner survived his job, he could earn his weight in gold and generate an esteemed reputation to pass on through generations.

When I first began this project, I did not know anything about mining. I had difficulty grasping the terms and processes being described. I am quite sure I asked a lot of stupid questions. I certainly saw a lot of faces cringe. However, possibly because of the current anti-mining climate in the United States and what appears to be the end of mining in Colorado, these men and women chose to tolerate my stumbling intrusion into their world in order to help document the end of an era. I have included a glossary in the back of this book for greenhorns like myself.

∞

Bill Rhoades. Photo by Rachael Pftenhauer

Many of the stories I heard during that first visit to the Hardrockers Holidays referred to a man named Bill Rhoades. Bill was clearly a legend, a mentor, and the source of many tales of both heroic and ridiculous behavior. At the following year's Hardrockers Holidays, I finally met Bill. A little man with a big cowboy hat, he sat in his pickup truck as people came to greet him and share a laugh. I was told that he'd recently been diagnosed with lung cancer.

Bill didn't want to talk to me. "Too shy," he said. But when he eventually got out of his truck to chat with the sheriff, I approached Bill again and tried to persuade him to grant me an interview.

The sheriff laughed and said, "That'll never happen."

Bill, frail and wobbly, shot the sheriff an, "Oh, so you think you know me?" look, then turned to me and said, "Next time you're in town, come by the VFW , where I work. We'll talk." Two weeks later I called the VFW to tell Bill I would be passing through Silverton the next day. The man who answered the phone said, "Bill Rhoades doesn't exist."

I was stunned by that phrase, especially when I realized it was the man's way of telling me Bill had passed away. For isn't that the fear of many people—that when they are gone, it will be as if they never existed? This may be especially true for miners, because at least in Colorado, they don't really have a next generation of miners to pass on the torch—the stories of their lives, proof they existed.

With Bill's passing, this project suddenly took on new importance and urgency. Not being able to include Bill's oral history in this collection is a great loss. However, I hope his presence is felt in the words of some of his friends and his son, Terry Rhoades.

The majority of the miners portrayed here were hardrock miners—mining gold, silver, lead, zinc, and copper—as opposed to soft metals like coal and uranium. Most of them worked in the San Juan Mountains, which was one of the most productive hardrock mining areas in the country, during a time when mining was a mighty force in the world. San Juan mines produced hundreds of millions of dollars in precious metals.

The San Juan mining boom began in the early 1870s. At first the sought-after bounty was gold, but by the late 1870s the local prospectors realized there were huge amounts of silver in the mountains. During the 1880s and the early 1890s enormous amounts of silver were shipped from the San Juans.

Prospectors were the front line of any mining exploration. They picked at the unforgiving mountains, looking for evidence of ore to "claim." It was a very harsh and lonely life. As one prospector wrote in his journal, "Here it is 1886. I am more than thirty years of age, and I am no better off than when I was only thirty minutes old." He died in an avalanche a month later.

Even when the prospectors discovered ore, the metal was still in veins locked into rock, requiring expensive forces of machinery and man power to extract it. Contrary to the tales and legends of many storytellers, most Colorado prospectors who discovered ore rarely trotted out of the mountains with saddlebags of nuggets, yelping and hollering about their newly found riches. Riches had to be mined out; excavated. Because most prospectors did not have the capital necessary to extract ore, they often sold their claims—for relatively small amounts of money—to wealthy business owners, rich investors, capitalists, and stock companies. Most of the capital came from the East Coast and Europe.

The new owners of these mining claims hired large numbers of workers to mine the ore. In the San Juans ore veins often follow a vertical slant. Drilling and excavating above one's head was dangerous and laborious. And support workers—including muckers, timbermen, nippers, and trammers—risked their lives daily. But so long as silver was in high demand, so were mining jobs. This was especially true for immigrants who were looking for a foot in the door to the "American Dream." Many men, and later a few women, jumped at the opportunity to make a good living doing something that didn't require an advanced education. It didn't matter how difficult or potentially deadly the occupation.

Mining accidents abounded—falling rocks, men tumbling into excavated chutes, collisions on the underground tracks. Without proper ventilation, fumes from blasting dynamite lingered in the air. Steam drills created a fine rock dust of razor sharp silica that impregnated the lungs. Thus came the term "miners' con," referring to consumption of the lungs—corrosive chemicals and the sharp dust eating through the lungs and throat linings.

It took a hard brand of people to work these mines, and communities held good miners in high esteem. In mining areas, life outside the mines was equally treacherous. Avalanches ran over mines, homes, roads, and people.

When silver prices began to slide during the 1880s, western silver interests influenced Congress to pass the Sherman Silver Purchase Act in 1890, in which the government agreed to purchase four and a half million ounces of silver each month. This protected the value of silver for a year, then the slide started again, and due to difficulties with domestic loans and international financial problems, Congress repealed the Sherman Act in 1893. The price of silver dropped dramatically.

The repealing of the Sherman Act would have killed the Colorado mining industry, if it depended only on silver, but gold (silver's more elusive sibling) remained valuable. Prospectors shifted gears and soon discovered increasing amounts of of gold-bearing veins in the San Juans. Old mine dumps were scoured to reclaim gold that had previously been discarded. And so began "gold fever."

Mining companies, interested in squeezing out profits, pressured miners to work long hours by making financial contracts with the miners. "Contract mining" guaranteed a certain amount of money per so much rock or ore they produced. Such an incentive plan also sometimes encouraged miners to work long hours and cut corners.

At snow-packed altitudes above 10,000 feet, men working such long hours in the cold and damp mines suffered pneumonia on a regular basis, and often developed arthritis. New machinery, powered by compressed air, introduced additional injuries when air hoses came loose. Relations between management and labor had been declining. Men who never before stole ore from mine owners, began to think about taking some of the precious gold for themselves.

Toward the middle of the twentieth century, the country's attitude toward mining began to change. Some say it was because of the fluctuation in ore prices. Some say it was because of environmental and mining safety regulations. Whatever the reason, one by one, mines across the San Juans began to shut down. Although there is still ore in the San Juans and a handful of small mines still operate, the Sunnyside Mine outside Silverton (also known as Standard Metals Mine) was the last survivor of the larger mines, and it ceased production in 1991.

The San Juan mining towns like Silverton, Telluride, Ouray, and Rico turned to other means of survival, mainly becoming tourism towns. The number of old-time miners is dwindling. In this book, they live on to tell of a time when these towns existed for the sole purpose of mining.

This book seeks to be a portrait of an incredible brand of individuals, working in incredible conditions, during an incredible time in the American West. Their stories reflect only a fraction of the tales swirling around the San Juan Mountains, like ghosts.

Verena Jacobson

Ouray, Colorado

I was born in Ouray in 1907, in the little blue house at the end of the ski hill. My father was an electrician in the Atlas Mine.

We had no television. There was no radio. Radio didn't come in until after World War I. They did have a picture show here, but no one was wealthy enough to go every night. You had to make your own entertainment. Us kids were content just making an impromptu picnic. We'd take wieners, buns, and potato chips. We did a lot of hiking.

I went into the Atlas Mine with my Dad one time. You get in this car, an ore car, and then they have this little motor. A man runs the little cars in and out. It gets perfectly black inside. They have lights every so far, and the man has a light on his motor. But then you just get to see the rock wall. That's all it is. It's just rock wall up above you and on either side.

The hard hats hadn't even come in yet. There were accidents every once in a while. If we saw a wagon coming down Jim Brown Hill earlier than usual, we knew that somebody had been hurt.

My Dad was an electrician in several of the mines: the Atlas, the Revenue, the Guadeloupe, the Black Girl, and the Bachelor. My uncle Frank, worked for my dad. Uncle Frank and I were very close. There were just ten years difference between us. He was my mother's youngest brother. So he was really like a big brother to me.

In my family we had only four kids, all girls. I was the oldest one, and when the two youngest girls were born, I was old enough to help in the house. Very few people had running water in their houses. You had a hydrant outside where you had to go out and get water and bring it into the house.

I was twelve years old when my mother started me doing the ironing. I just stood on the floor and used the kitchen table for the board. This lady—the midwife who came to help my mother when my youngest sister was being born—was shocked that I could iron. She said her daughter was in her thirties and she couldn't iron. Apparently her daughter was a beautiful violinist. I think her daughter was real smart to keep practicing her violin long enough so she didn't have to help with the housework.

When WWI came along, Frank went into the service right away. During the war, the 1918 flu epidemic hit Ouray hard. It was really bad. They didn't have antibiotics. We had the hospital then, but the doctors and nurses didn't know anything about influenza. And many victims never even made it to the hospital. A lot of people died.

The town health doctor proclaimed that there could be no public gatherings, not even outside. This was right when the Elks were preparing their Christmas program, where they handed out candy to the kids. But the doctor said, "I don't care what you do with your candy. There's not going to be any public gathering."

The Elk members got together, and they wondered if maybe Johnny—who was the man that ran one of the horse barns—would lend them his sleds and horses. Johnny did and the Elks went around, house to house, handing out candy. This has become a tradition now.

When Frank came back from the war, he worked pick and shovel mostly. Then he started working for a druggist in town. Eventually Frank and Albert Schnieder bought the drugstore. I was in my teens. I worked for them down at their store. The post office was also a part of it, so they had to open up at 7:00 in the morning, and they couldn't close until 10:00 at night. The store had to stay open so people could go in and get their mail.

When I graduated from high school, there were only five of us in my class. Only one boy. I never was too awfully interested in boys. After I graduated, I taught school for some kids at the Yankee Boy Mine, up above the Camp Bird Mine. There were a lot of mines up in there. The miners worked seven days a week. They never got time off, except for a long change of shift. But most of them stayed up at the mines.

A lot of avalanches would come down. One slide ran and took out the Camp Bird Mill and part of the Camp Bird Mine itself. Another one of my uncles worked in the mill. They didn't find him for twenty-four hours, but he was still alive. He happened to get caught under a lot of machinery in the mill. That sort of formed a little hood over him, so he had air. He just kept calling and calling until they came and got him out.

I guess you could almost say avalanches were the biggest fear. But you know, we never heard much about fear. I guess we just accepted it as it was, and let it go at that. Nobody ever got very excited about anything. People just took things as matter of fact. We didn't have TV, which drums and drums forever on the same subject, whether you want to hear it or not.

My sister got married and moved away. The following spring, she was going to have a baby. She asked me if I'd come help her out because she was living on a farm and it was hard work. I said I'd come and help for the summer. So that's what I did. Of course neither of us knew "sickum" about farming. Not a thing.

My sister's husband had a brother. He started taking me to dances, and to a show now and then. And it ended up that we got married.

After my husband died, I came back to Ouray. I came back to take care of Uncle Frank. He was getting old and feeble.

Frank was on the council. He told me they were having problems because somebody wanted to start a liquor store. The whole deal made him laugh, because there was this one other guy on the council who was a very anti-liquor man. Tremendously anti. And this guy argued, "Oh I don't think we can do that. We already have one liquor store." It didn't stay that way for long, of course.

I worked over at the museum a lot. The gal that was the curator then, her name was Vera. One time we were talking about the St. Elmo Hotel. There's a room number thirteen at the hotel. A lady we had known, named Kitty, had a son who was supposed to have been shot in that room. Kitty, herself, had since passed away.

Vera was telling me, "You know they say they have a hard time with room number thirteen, because there's a ghost." Somebody had told Vera that people would wake up in the middle of the night in that room, and a lady would be sitting in the rocking chair, smiling and rocking.

I said, "Well, that's stupid."

She said, "Well, you know, somebody said it was Kitty."

I said, "Oh no. If she was smiling, it wasn't Kitty. She never smiled a day in her life!"

Vera said, "Well I guess I can put that one to rest."

And I said, "Yup, put that one to rest."

We have a lot of new people in town. A lot of people come here because they think it's so pristine and so beautiful. They get here and decide to stay. Then they decide that they want it to be just like Los Angeles where they came from. And then they get busy.

Of course I don't know a lot of people because I haven't gone out so much. Uncle Frank was an invalid toward the end, so I stayed home with him. He was very alert, mentally. He just couldn't walk. His knees and, eventually, his hips gave out.

One Christmas, right before Frank died, the Elks were short on cash to pay for treats for the kids. Frank heard about it and so he saw to it that they had enough money for the candy. He was well-to-do.

Frank and I were having coffee one morning. I got up and turned to the sink, and I just sort of felt him walking out the door. When I turned back around, he was gone. He had passed away.

Albert Fedel

Ouray, Colorado

I was born in Ouray in 1919. My father came from Austria. He moved to this country to mine. He first went to Pennsylvania, because a lot of his people who'd migrated were working in the coal mines there. These guys had always been hard-rock miners, not coal miners. So when the gold bust came along, they moved west to the camps where the rock mining was thriving—Cripple Creek and Leadville. Eventually they came over to Telluride, Silverton, and Ouray.

They were tough men. Good people. Talk about camaraderie. During my time as a kid, if old Joe didn't show up, Boy everybody in the mine knew it, and they had to go find out what was wrong with him. People were very close.

We had five grocery stores and about twenty-eight bars. When the miners came to town, there was a lot of drinking. Very few disputes. A couple of shooting messes, but that's to be expected. If the city marshal found drunk miners out on the street, he'd just take them and put them in the pokey to dry out.

There was a lady named Mary. A miner was going to marry her, but he never showed up. Mary went about half cuckoo. Went off her rocker. She stayed living in an old boarding house even after it was vacated. The windows were all broken and knocked out. Her sister, who we called "Old Minnie," would make sure Mary had food and everything else. In the morning, Mary would get up and build a fire on top of the stove. The smoke would run all the pigeons off. They'd come up flying.

Down at the end of Eighth Avenue, there used to be a bar. They had a still under the sidewalk. The guys, teamsters, would come down with their wagons and dump the ore off at the railroad. There was a creek and a swinging bridge that went across to the still. They'd drive their wagons into the river to cool off the wheels and soak the wooden spokes. We'd hold the horse team for them while they'd go in and have a couple shots of bootleg. Then they'd come out and go up to the barn with the team. You could hear the music going twenty-four hours a day, due to all the bars and dancing and gambling. But it wasn't nauseating. It didn't bother me much.

In the old Western Hotel, Mary, the Mrs., used to have lots of miners there. She was sure a nice old gal. She'd feed any of them. Get them a job too. And if a guy didn't show up for work at the mine, the boss would call down and see if Pete, or whoever, was there. And if he happened to be there, she'd drag him out of the bar, put him on a horse, hit it on the rear end, and away he'd go.

They were good miners. The bosses didn't want to lose them. When the miners would get paid, they'd come down to town and pay her off.

I started working in the mine, during the summers, when I was about fifteen years old. When I got out of high school in 1937, I went to work at the Treasury Tunnel. It's now known as the Idarado Mine. We were on the first development of that mine. During the Depression days, ore was taken out of the mine by mules.

The first time I went into a mine, it was exactly as I expected, wet and dark. We used carbide lights. The ventilation in a lot of places wasn't good. There were problems with bad air. We always had candles going out. Boy, when that candle went out, you were ready to tramp. Get out of there fast.

In the 1930s they used to have lots of fatalities. When you're running down through the vein, even on a drift, you'll hit a spot where the ground is very bad. Doesn't hold. These days you can take a stoper and you can drill all kinds of holes, screen it, and cement it with special steel bolts placed in the drill holes. It'll hold with the steel timber there. But years ago, we only did old-fashioned timbering. You had to be an expert or it wouldn't hold.

When we had the bunkhouses going up there, you went up and stayed until they said you could go to town. It's not very far from the Idarado to town. But the road wasn't much more than five feet wide, and it wasn't paved. I went up in September, one time, and I didn't get back until July. We were contract mining, which is where you get paid according to how much rock you break. We made good money, but we didn't have the chance to go to town and spend it.

They had these outhouses, until the bunkhouses got plumbing. And you always got a guy that would swipe the shower soap. We couldn't catch this one guy, so we finally decided we'd fix him up. I went over to the blacksmith and got a couple of Gillette razor blades. We powdered them up and put them in a bar of soap and left it in the fountain.

He looked like hell when he got done with that bar of soap. Oh man was he cut up. Mainly the arm where he started. He come on out yelling and headed for the medical box. The boss said, "Well I guess you won't never bother another bar of soap." He handed him his check and said, "Now you can go find another place to work."

∞

In 1940 I went to Western State College. I was there two years. Then I got my pilot's license. The war was coming on, so like all young bucks, we decided we were going to win it. We had a fine tour there for a number of years.

When I came back, I went to the Camp Bird Mine. Some of those old miners, you couldn't get along with them. Those old Swedes and Polacks, they were ornery. When they'd come out of the bunkhouse at night, you'd hear them arguing with the trammers. The trammers would be catching hell because they'd went off the track some place and didn't get the steel to the miners or something. There were also some pretty strong arguments around the pool table. Mainly they'd argue over who had run off with their high-grade, the gold they hid for themselves.

There was a lot of high-grading. Management caught on and when you'd come up in the dry room, they'd make you strip. But the miners would pull some tricks. Put the ore in a thermos bottle. When you went outside and they ran the check, you were clear. Gold was twenty dollars an ounce; that was something good. They finally caught on to the tricks too. You usually got canned for high-grading, but now and then, guys would get away with stuff.

I did a lot of stoping in the Camp Bird. I didn't like it. Stoping is real dangerous. You got a hose tied on to it and a machine runs off the cylinder, rotating. You ever hit a hang-up and you're not real quick to cut that air off, why you'll get wrapped up in the hose and everything else.

At the Camp Bird, they mined it right into where you could see the roots of the trees. They didn't leave anything at all.

I think Red Mountain Pass has more avalanches than anywhere I know of. We had an avalanche group in town. I went up on that crew to help find two guys that were buried in a slide. When we finally found this one young man, he was frozen in the same place he was running. If he'd gotten another ten feet, he'd have made it. The other guy was under a piece of timber and he survived. He had a pocket of air.

Those slides are mean, Boy. They sure make a rumble. And that force in front of it—you see those trees and everything coming up yards and yards ahead of the body of snow, just from the force of it. In split seconds, it's down there.

Once they had a slide that came over the portal to the mine. It came and took out the snowshed and the compressor room. Everything shut down, the lights and stuff underground. I was on the outside. We finally got an air line in. The guys inside got out through an escape exit about a thousand feet in from the tunnel entry.

We had operating mines in Ouray all the way up until the 1980s. Your base metal mining was still relevant, due to the fact of the military. Then environmental regulations started the layoff period, and the mines started closing down.

When I returned from World War II, I had an opportunity to buy property in Ouray. I went into a large gas, oil, and auto repair business. And I spent from 1957 to 1979 as mayor and a councilman.

All the old-timers left. But a lot of them, who never sold their property, they've come back here to live out their retirement years. It's funny that they would return.

The devastating remains of a snowslide.

John Crim

Ouray, Colorado

I went into the Alta Mine, near Telluride, in 1935. I was fifteen years old. I had lost my father when I was fourteen, so I figured it was time for me to kind of help out a little. Mining paid bigger wages than anything else. I started out mucking for $3.75 a day. It was cold, wet, and dark. There were other people nearby and it didn't scare them, so I figured it shouldn't scare me either. Big macho kid and all that.

But there was one incident I was very fortunate to get out of. I was then mining in the Idarado. I had drilled out the beam above the muck pile. I went down for dinner and came back with two fifty-pound sacks of dynamite powder. When I walked out under the area I had drilled, I heard something. I don't know what I heard, but I threw that powder down and ran like crazy. A solid chunk, about ten feet wide and thirty feet long, fell right at my heels. That would have just squashed me like a bug. And there was no reason why the powder didn't go off when that big chunk fell on it. It should have. I sat there and said a few "Hail Marys." Then I got some more powder and went back to work.

I know several times when a few of us got away with things like that. Maybe miners got some sixth sense or something. They call it "tommyknockers." Tommyknockers are supposed to be these little guys, rascals, that blow out candles when you're not looking. But they also warn you with strange noises when the mine's going to cave in, and things like that. I'm pretty sure the way it got started was like in my case, when that chunk started to slip, it made a funny grinding noise and I knew that something was going to happen. Anytime the tommyknockers start squawking at you, you better do something about it.

I don't know how I became a boss. One time the mine superintendent came through. I was drilling some holes, getting ready to blast, and he shut my air off. The machine wouldn't run. He said, "You like working as a miner?"

I said, "No."

He said, "How would you like to be a shift boss?"

I said, "Well I'm not very crazy about that either."

He said, "Well I'd sure like you to run a crew for me." What I understood about the job was that you got good money regardless of production. So I told him, "Yeah, I'll give it a try." And that's the way I got to be a supervisor in the mine.

I had probably a total of thirteen years of experience. It was a steady paycheck, but I really didn't like it too well. I had to tell guys, who had more experience than I did, what the hell they had to do. They kind of looked down their noses at me, and I don't blame them. Here's some punk kid coming along and telling them what they should do.

One time I was running a crew for the Camp Bird Mine. I sent the crew underground one morning. I followed them in, as I usually do, to find out what they were doing and if it was satisfactory. But pretty soon I realized I was in dire need of a "facility." And I knew I was going to have to hurry to make it outside to the outhouse.

I got to the portal, the opening to the mine. I threw off my hat and my rubber coat. I dropped my lamp belt and scurried as fast as I could to the outhouse, and I made it. But then I heard this buzz. I thought, "God, the flies are bad in here. I've got to get some quick-lime and get rid of these flies." Then I looked up to the top of the outhouse and there's this great, humongous hornets' nest hanging down up there. About that time one of them hit me on the ol' bald head. So I thought, "Well I better get out of here," and I threw the door open and I scurried up through the choke cherry bushes and oak brush. That was the only time I ever streaked. I don't think there was anybody who saw me do it. They would have gotten a big kick out of that.

I was running a crew in the Idarado Mine one night, and I saw wood smoke coming down the drift. Over at the Pride of the West Mine, some seventeen men were killed when they had a fire outside and the smoke and gasses went underground and suffocated them. So seeing that smoke scared the daylights out of me. I grabbed two fire extinguishers and went checking all the little raises that came up from below and finally found the one the smoke was coming out of. I got on the skip and went down to that level and crawled on my belly back through the drift and, sure enough, there was a slusher platform on fire.

I got the fire out and I went back up in the stopes where there were six people working. I said, "You guys smell any smoke?"

They said, "Yeah."

"Well why didn't you get the hell out of here?!"

"We didn't think it was very bad."

I said, "Well it could have been real bad."

If the rest of that platform had burned up, why I imagine we might have lost some of them. There was a lot of pressure to keep working. Most of the miners, they worked by piece work—so much a ton or so much a cubic foot—and every measuring day you got paid on that amount.

Then they'd usually go to town and get "half-cranked up." Of course I'd have to go too, sometimes. After so many drinks, most got to thinking they were really good looking and rich and could whip anybody in town. And some of them were pretty good at it.

We used to catch the bus, to go up to the mine, in the middle of town. One night—I don't know why I was driving the bus that particular night, but I was—and I saw poor old Pete come staggering out of the bar. And I could tell he knew where the bus was, all right, he just couldn't catch it. It kept moving around. He finally got to the door, and he started to get on the bus.

I said, "Pete, I can't let you go to work like that." Oh, he puffed up like a poisoned toad. And I said, "Well, Pete, I can't." So he climbed off the bus, took his lunch bucket and went right back over to the bar.

The next night when he came to work, I got up to where he was working and I said, "Pete, what did Gracie say when you went home last night?"

He said, "Well I went home about 2:00 o'clock in the morning and I couldn't make it to the door, so I crawled on the floor and I banged on the screen door. Gracie came out there, and she looked down at me, and she said, 'Pete, what's the matter with you?' I looked up at Gracie and said, 'Oh, Gracie, that Idarado Mine is so gassy!' Of course Gracie didn't fall for it. She knew what was the matter with Pete. She'd lived with him a good many years.

Right after the war, when the guys first started coming back, the town got pretty rowdy. So the city hired an ex-Texas Ranger to quiet the town down. There was one guy who was kind of ornery. We called him "Bungaloo." Well we were at a dance down in the old Pavilion one night, and I saw Bungaloo sneaking up behind the Texas Ranger out on the edge of the curb. Bungaloo binged him one right on the jaw and the Texas Ranger fell off in the gutter.

Bungaloo ran down the alley, but there was a clothesline there, hanging down. I can still see Bungaloo's feet sticking up in the air when he hit that clothesline. Must have caught him right under the chin. But he got himself together and took off. The Texas Ranger lasted about three or four days after that. He realized he couldn't handle it. The guys were just too ornery.

After that they conned the worst one of the bunch into being the Town Marshal. Letch Clark. It was smart thinking, but a letdown for old Letch. He was probably one of the worst to get into it downtown, but he couldn't while he was law enforcement. So the town settled down pretty good.

My wife's first husband and I were partners in a rock shop here in town. Then he got sick with lung cancer. He had worked in the uranium mines for quite a while, and smoked cigarettes. A year or so after he passed away, Helen and I were married. That was thirty years ago.

My health held up reasonably well. Most people that worked in the mines for very long got some silicosis on their lungs. That's something you never get rid of. It just keeps cutting. And you get big scars on your lungs, and your oxygen isn't easily converted to your blood stream. If you got it bad enough that you had to leave the mine, you sometimes got a little state compensation, but it didn't happen very often. Most of the miners just worried through it until it finally killed them.

I think what probably kept miners going back again and again was the challenge of being able to make advances under bad conditions. And they always figured, "Someday I'll find the perfect piece of high-grade," you know, stolen gold specimens. Some guys put it in their boots and about every place you can think of. When the bosses started searching them as they came out, the miners would stick specimens in their candle cans, which they'd hold up with their arms when frisked.

Once in a while there would be a little trouble with high-grading. People got to taking too much time trying to get a little gold, and they weren't doing enough work. But usually you could just say, "Hey, look, go ahead and take a specimen, but get your work done before you do it." And that worked pretty well. They were hard workers.

Mining was a way of life. You went to work and hopefully you got back the next night or the next day. I don't think Helen ever had any qualms about it. Of course there were a few things I didn't tell her either, or she would have been uneasy, I suspect.

I have three boys. I told them they couldn't mine. It's a lot of hard knocks and you got your neck sticking out a mile sometimes. I just didn't want to see them do that.

Frank Hitti

Dolores, Colorado

I don't remember my real Dad too well. He got gassed in France during World War I. We lost him not long after he came back from the war. My mother got married again in 1929, the year of the Great Depression. I was nine years old.

My stepfather was a miner. He was a foreman up at the old Sunnyside Mine in Silverton. During the Depression, all the mines that were going in Silverton collapsed. That was it in Silverton, except for the Shenandoah-Dives Mining Company. They started up the Mayflower Mine. They could get cheap labor. My stepfather went to work for them.

I was never interested in going into mining. I was slated to go to college. I did go for two years, but the finances ran out, so I had to come back and work. That was about 1940. I joined my stepfather in the Highland Mary Mine, north of Silverton.

At that time there was no such thing as being able to just go in and get right into mining. You had to start as a miner's helper, a nipper, for three-to-five years. You go in and you help the miners set up their drill machines and so forth.

I looked up to the miners. They were kings. I admired them. "Christmas Tree" John was one of them. John was driving drift. The guys on the shift before him, they had blasted, but there was a hole that still had dynamite in it. It hadn't gone off. Christmas Tree John was an experienced miner, of course, but he didn't clean out enough before he drilled his hole this time. He hit that dynamite with his drill and there it went. He was blown to pieces.

I don't know why he was called "Christmas Tree." Maybe he was always lit. That's very possible. It's a tradition. That was my tradition too. It seems that needing that drink at the end of the day was what started most miners' drinking problems, other professions too. You needed to relax. You came out of the hole, and you have all that stress that you've gone through all shift. A double shot and a beer is the first cure. And then, finally, it's every day, except maybe on weekends or holidays, or maybe then it's more. We were hard working, but we had to party too.

I had my share of accidents. The worst one was in Idarado. A slab came down on me. It landed right square on my head. My partner took me out. I was unconscious almost three hours. When I did come to, the mortician was bending over me. That was kind of funny when I look back on it, but it didn't set very good with me at the time. The mortician said, "Why, I just come up with the Doc, that's all." They finally took me to the hospital in Denver. They operated and did a three-vertebrae fusion. They took out part of the bone in my hip and put it in there. It came out all right.

I've been lucky. Three or four of my friends were killed in the mines. The last one, he was just walking by a place. He was going to put timber up, to secure it. He was looking at it and a big slab—about eight feet long and one foot thick—fell on him and squashed him.

There were others, like "the Jew." I don't know if he was Jewish or not. That was just his tramp name. A tramp miner would work ten days, get his bills paid and have enough to go somewhere else. Anyway, he was working up there and lighting the fuses. Of course we didn't have these electric lights in those days, just carbide lamps. All we can figure out is that he was working by himself and his lamp must have gone out. In the darkness, you couldn't find where you were going. In the meantime the fuses he had lit burned down and the shots started in. The blasted rock covered him up. He was a mess. There wasn't much left of him.

We played practical jokes in the mine, to sort of cut the edge. A partner of mine, who used to work with me up at the Mayflower, he was deathly afraid of mice. Johnny was his name. He used to be a fighter, an amateur Golden Gloves man. Of course the dog house, where you go to eat lunch and so forth, had a lot of mice in there. One time we caught a mouse and I don't know where Johnny was, but before he got there, we put the mouse in his lunch bucket.

We were there eating and waiting for Johnny to open that lunch bucket and, my God, when he did, that mouse jumped up. Johnny went absolutely wild. He cleared out of there, and he went home. He'd have killed us, he was so mad. After that he laughed a lot of times about it too. But, you know, I lived right across the alley from him in Silverton, and I didn't know he was that scared of mice. Not like that. Somebody obviously did know it, and so they pulled that on him. But they never did that again! Never did—no, no.

And then there were different things about lunch buckets. After you ate lunch, somebody might nail your lunch bucket to the bench where you sit. At quitting time, why you go out and grab your lunch bucket to go home. One time they nailed mine down to the bench. I grabbed that bucket and the lid came off, and I didn't have a lunch bucket no more. They thought that was pretty funny, but nobody admitted who did it.

There was a lot of joking, but in the bars, sometimes the joking would turn to fighting. It would start off friendly. Then you'd say something wrong to a friend of yours, or an acquaintance, and he'd pick it up. "I can do this better than you can," sort of thing, and it went from there. We'd go outside and settle it, have a little fight. One guy gets knocked down or maybe both of you. Or, maybe, you'd just shake hands and have another drink.

The bars could get pretty crazy. One time this guy, Ben, and a bunch of us got together in Silverton at what they now call the Club Café. We all got pretty well "soused-up." At some point Ben disappeared. I didn't know where he went. When he came back, he was riding his pinto stallion. Here comes old Ben on that big black and white stallion, and he's a'whoopin' and a'hollerin', and he came in the door on his horse. That Club Café was kind of a big hangout. It had a little restaurant, and there was a dance floor. Everybody got in on chasing out Ben and his horse. But it was the horse that got him out of there.

∞

All of my buddies are gone. I haven't got a person left that I worked with. A miner's life isn't very long. You get silicosis from breathing that dust and so forth. And accidents. I've got arthritis real bad. Arthritis is one of the most common problems with miners. That arthritis hit me so hard I had to have a complete hip replacement, and a complete shoulder replacement.

But silicosis is the worst, because of the silica cutting your lungs all to pieces. You got to get supplemental oxygen, and you're susceptible to hemorrhages. You just get so weak. Silicosis is what a lot of my buddies have died from. Of course most of them were also heavy smokers and drinkers too.

I'm a teetotaler now. I haven't had a drink since 1974. I had to quit. I had become a slave to it. I just about lost everything. I lost a lot of money, almost lost a wife. I was drunk all the time and then sick from drinking. Pretty close to cirrhosis.

I spent about two months in the hospital. My liver and then the gout hit. Then I caught double pneumonia in the hospital. Dr. Parker, who was with the Telluride Mines—I knew him quite well—he came and saw me. He said, "Frank, you've about had it."

I said, "Yeah, I know." Of course, I didn't know anything, they had me all doped up on morphine and other drugs. They called all the family, my kids. I knew something was up. But I made it. When I went home, my family had taken all the liquor out of the house. I haven't had a drink since. And I don't miss it either.

There was something else that helped me make that transition. After I came back from the hospital, I spent three months down in Elk Park with just my horse, my dog and myself. Elk Park is south of Silverton. I stayed in an old boxcar, and I prospected. The boxcar, I imagine, was kind of a safety deal for hikers. It had a little kitchen in it, a pot bellied stove and two bunks.

The motor men on the railroad would check on me every now and then. They would stop and bring me something to eat. There were two little rail cars—one runs before the train, and one after. The one that goes ahead of the train, checks to make sure the tracks are clear for the engine. The one after makes sure that there aren't fires from the steam engine. That guy running the rail car after the train had a little more time, but the guy in front sometimes stopped too. He was about thirty minutes ahead of the train, then he'd have to go.

It was hard on my wife and the children, because I didn't come home for three months. But it was something I had to do.

That was the last real mining I ever did. I loved to prospect. I picked out most of those mountains above Silverton. My prospects are still up there. Nobody else can get to them.

Barbara Spencer

Ouray, Colorado

We came to Ouray in 1921, when I was five years old. My father, William McCullough, was one of the first licensed mining engineers in Colorado, number 169. People have said that my dad was a bit of a hero. He would say, "You just did what you could."

The road to Silverton wasn't paved, so it was closed in the winter. The first car over in the springtime got its picture in the paper.

But there was that terrible winter when Silverton got so snowed-in they couldn't get any food or supplies. When the situation got desperate, my father and his partner, who were working up at the Lucky 20, went over on snowshoes with food and supplies. His partner was a tall man, and my dad said trying to keep up with a long-legged fellow when you're short and squat, at 10,000 feet, was quite a chore. But they got the emergency supplies there. And, eventually, the *Denver Post* sent in a plane to drop the other things the town needed. Of course the first thing they dropped were bundles of newspapers.

I remember, one Christmas, my dad was saddling up his horse to take some things to one of the mines. He loaded his saddlebags with mail and groceries. And the last thing he picked up was a box for the cook and his wife and children. My dad put the string of the box around the saddle horn and it said, "Ma-ma! Ma-ma!" He was so startled, he fell back in the snow. He didn't realize it was just a doll.

Those were Depression days, the days when folks had to look after each other. If you were prospecting, the grocer would carry you for months on end until your shipment went out and you got a check back from the smelter. But you know they say a mine is a hole in the ground with a liar on top.

I remember one young man ran up a sizable bill at the post office/drug store—that's where you got medications and school books. And this young man went out and bought a new car anyway. That car went down Main Street and Albert, one of the store owners, said, "Well, there goes the bill. Guess that's part our car."

The fellows that ran the post office/drug store, also ran the movie house. There was a lady who accompanied the movies on the piano. When it got scary she would play scary music. And when there was a march, she'd play right along. The first movie we went to was "The Three Little Pigs"—a cartoon. My dad just laughed and laughed, watching the wolf try to blow the little pigs' house down. He laughed until his face turned purple.

The local newspaper once ran a very nice article about my father, and that's how I caught my husband. I went down to get extra papers, and I thought one of the editors had such big blue eyes. He had a brother and both boys dated me.

Then, one time, I was cross-country skiing and I was at the "drinking cup," which is a spot where water trickles out of the hillside. A snowslide started running off the cliff in front of me. I could hardly ski over it, and it was storming hard. My mother sent the newspaper brothers to come look for me. And so I married one, the one with the big blue eyes.

My husband was also a prospector when I met him. He had a little donkey to carry his supplies up and bring ore back down. When I married him, we went to Oak Creek to run a newspaper, so he gave the donkey to my dad. A friend of ours said that my dad got the better of the bargain! Well, eventually, my husband decided, "The heck with it. Let's go mining." So I came home. Mining was home.

Everybody who came here back then, came here to mine. There were a lot of foreigners. The northern Italians really knew mining. They swept into the country. I remember hearing Italian women gossiping on the street corner, and I couldn't understand a word.

∞

The winters were hard and brought everyone together. People were killed and killed in avalanches. The men from town, who were able, would take tall, steel iron poles and probe along the avalanche looking for survivors or bodies. My husband went out on the probe line when Reverend Hudson went over to hold Sunday services in Silverton with his two little girls, and that avalanche slide killed them.

Sometimes people survived avalanches. One time a guy got caught in a slide, and he landed down at the creek. He was presumed dead, but there was enough air between the snow and the water that he was able to breathe. He made it up to an old tollhouse on the road. A couple of guys were there; and when they saw this frozen man, they ran because they thought he was a ghost.

The worst was that big avalanche that ran and killed some miners. My husband went up with the other fellows before daylight, probing. I'll never forget his description of the pitiful procession of women, the widows of the miners killed in the avalanche, and their children, coming down over the avalanche in the blue light of early morning. It still makes me cry to think of it. They were coming to town to be looked after.

Barbara (standing, on right) and her family. Courtesy Barbara Spencer.

Everyone had a group who would take care of you if something happened. If you were Italian, the Italians were around you. If you belonged to a certain church, that church would be around you. If you were an Elk, the Elk brothers would be there.

The Elks were a very important group here. They had a dance every month and you dressed for it. You wore long gowns. But there isn't so much happening with the clubs anymore, not since the advent of television. And Ouray has changed quite a lot since the mines closed. There used to be twenty-seven saloons on Main Street, with that sour old saloon smell coming out of them.

I remember when they went to rehab Lake Emma that caved in that Silverton mine. There were whiskey bottles all over the bottom of it. You see, those guys were not going to carry a bottle around where the boss could see it, so when they were through drinking, they'd just throw the bottle in the lake. And the bottles sure accumulated. But I heard you can find the best bottles in the old privies. Miners would go in there and have a nip out of a whiskey bottle then drop it in the hole.

Camp Bird Mine, Upper Level, 1930. Courtesy Barbara Spencer.

∞

The only mine left here is the Grizzly Bear. My son worked there. One time he was taking a load of ore from the Grizzly Bear down to the loading platform. His engine quit. He tried to start it again, but when he got to the edge of the cliff, he jumped. The truck went down and hit a tree so hard it snapped the top of the tree and kept going. When my son started to stand, he realized he'd broken his leg. He said that he sat there leaning against that bank thinking, "Lord, how come this happened to me?"

And the Lord said, and I repeated, "Because your engine quit." My son left the Grizzly Bear because they didn't offer health insurance, and he decided he needed it.

In my husband's day, none of the mines provided health insurance and such. When a rock dropped on him from the top of a 700-foot shaft, and broke his arm, that was it. He had to retire.

But it was hard for me when the mines closed. Even after my husband passed away. I guess I'm just a natural born Luddite.

I'm still trying to peel my fingers off the mines. I would never consider leaving here. I just pull these mountains around my shoulders and I'm very comfortable.

Old miner in Ouray named Buck LeClerc, circa 1970. Courtesy Barbara Spencer.

Buck LeClerc sharpening drill steel, circa 1970. Courtesy Barbara Spencer.

George Cappis

Telluride, Colorado

I went to work at the mines in Telluride in 1941. I was eighteen. There were three different mines then that are all Idarado mines now. I hauled timber underground, unloaded it and came right back out.

Then they talked me into going into a drift. First time I rode in, it was really scary. I'd been underground a little bit, but not to work. I only lasted about ten days in that drift. It was real dark and wet. It just didn't appeal to me. So I went to work one morning and I told the boss, "I quit."

I saw him that night. He said, "You didn't have to quit. Come back in the morning and rustle timber." I worked on that for maybe a couple more months, and then he asked me if I wanted to go on the mainline haulage—the muck train. I told him I'd give it a try. I must have liked it, because I worked on it then for about twenty-six years.

Aside from the muck trains and trams, they had four-wheel bicycles we rode in the mines. They'd go on the rails. Oh, you could really get around that mine in a hurry with them, instead of walking or waiting for a motor. The Idarado went clear across between the Red Mountain side and the Telluride side. Our mine superintendent, from the Red Mountain side, he'd travel on bicycle. He would come through there, down two raises and outside here in thirty minutes. Boy, you could really travel on them things.

One time the boss from New York came in, and he wanted to go back where they were drilling. So they loaned him my bicycle to go back there. He had a couple of wrecks on it before they returned, because the bike wouldn't stay on the rail. When they got back, he wanted to know whose bicycle it was. They told him and he said, "You get him a brand new one." He didn't much like my bicycle.

Mining changed a bit while I was there. At first we had mostly old, run-down equipment that wasn't very good. Then they started improving. They bought new ore cars to haul the ore out of the mine. They got new drilling machines for the miners. Things were better. But people still got hurt lots of times.

But what I remember more than people getting hurt, were people getting killed. From the time I went to work there until they shut the mine down, I think there were twenty-two fatalities on just the Telluride side. I don't know how many were on the Red Mountain side. Probably just as many. A lot of times it was fallen rock, but I think the majority of them were guys falling down raises and ore passes.

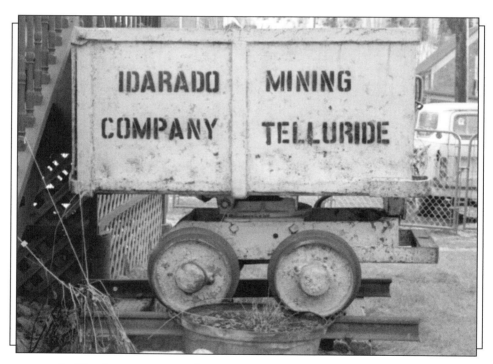

An Idarado ore car—a very big souvenir.

The last two guys that I had to help get out were the hardest ones for me. The first guy, when we went to pick him up, he was dead and still looking up in the raise. He had been timbering. The block, which the cable goes through on the post, came loose and hit him right in the face. They hadn't put the block on good timber. So I had to pick him up and get him out of the mine. But I think what got me worse was that because all the other higher-up bosses were out of town, I had to go tell his wife.

That was the only one I ever had to do that with, tell the wife. It wasn't very good. I took the next door neighbors with me. The miner's wife was sitting on the floor, reading. She had a little baby on her lap. Of course she knew something was wrong, right away. It was terrible.

Then there was that time when two guys were putting new rail across a raise. As usual they used a board to carry it across. The first guy got across with his end of the rail, but the board broke under the second guy and down he went. They couldn't get him out on top, so we had to pull him out of the bottom. I had to pull him out.

Every time I'd go back there, I'd have to keep looking over my shoulders. The area just spooked me. Some guys call the ghosts in the mine "white boots," because there was a time when the only good hard-toed mining boots they could get us were made in white.

I didn't like mining itself. Something about them slabs over my head. There were a couple of times I was really scared. Once when the ore pass was hung up, we were blasting and couldn't get it. The hang-up was about 185 feet above us.

We were working between shifts, the three of the big bosses—the mine superintendent, general manager, and mine foreman—and myself. I was sitting in the chute with this great big light, watching the pipe as they put it up. We'd go up eighty-five feet and blast it, and wouldn't get anything. So, I thought, put another pipe on. That pipe just bent. So we went up 120 feet and hit something. We pulled thirty-five sticks of powder up there and blasted it. A huge slab came down. Took that chute and everything out. I wondered what would have happened if the pipe had hit that slab and it had come loose. If it had come loose when we were under there with the pipe, there would have been no way anyone of us could have gotten out of the way.

Another time we were driving drift, that mill-level drift. There were only about ten of us working. We saw this slab above the air lines. So we thought, we'll go get some boards and put them up there and bar it down. We were doing that, and as I reached over to get the shovel, a little rock hit me on the hand. Then a big slab fell. When it hit the air line, that air line broke and struck me on the shoulder, knocking me out of the way of the falling slab.

The air line tore my light completely off of my hat, tore it all to pieces, and it bound me in a ditch of water. I just about drowned. Everybody had run for their lives. There was no light. But pretty soon somebody came back and shined enough light where I could see to get out of there.

∞

A Black Bear tram and a "liberty bell."

We also had our fair share of avalanches. There was a house out there by the mine. Two women and a little baby were in there and a snowslide came down and took that house away. It killed the two women. The little baby was in a box behind the stove and lived.

When the first woman went to work at the mine, some of them old, rough, hardrock miners, they didn't think it was a place for a woman. As far as I know there was only one woman who actually mined. Most of them were just nippers. They hoisted men up in the stopes and sent up the supplies. But those miners gave them a bad time. The language they'd use—they just wouldn't treat them very good. But once the women got into the workplace, the guys finally got used to them and that was the end of that.

But Telluride was still a pretty tough town. There was a guy named Snodgrass, I think. Some guys came into town, and they wanted him to take them up to Deep Creek to maybe buy some of his claims he had up there. I think this was in 1950. He was supposed to have had a money belt with him when they left—supposed to have had quite a bit of money on him. Nobody's seen him since. His family still comes up looking for his bones. Those people he was with said that something happened to him, and they had to leave him by a fire and come down and get help. But they never could find him or his remains.

There were times when there were a lot of fights. Miners, they just didn't take nothing off of nobody. Paydays were when the bars would get rowdy. The night shift would go out and get their checks, and a lot of them wouldn't make it to work that night. But most of us married guys wouldn't stay in the bars that long. Maybe go have a beer or two, then go to work.

We had some characters that worked for the mine. Joe and Tito; they were Italians, and they made their own red wine. They brought a thermos of that wine to work every day and guarded it. You could smell them fellas, the smell of their wine, coming up the drift. You knew who was coming.

There were a lot of foreigners in town then: Finns, Swedes, and Italians. They seemed to get along. They all had their own part of town they lived in. They had their own little halls, where they had their own doings and stuff.

In 1983 they started shutting down the mines—started the reclamation. Started getting rid of everybody. But they had to keep it open so the water could keep coming out. I went in there running the hoist one day and the cable broke. I was making a trial run. Joe and Cecil were going to go up and inspect the next one, and I had it about half-way up.

I put it on the full throttle; and, boy, it came to a screeching halt and that cable came down the raise. That's the last time I ran the hoist. The law says you have to not be over sixty-five, and I was. So they didn't insist that I run the hoist after that.

My wife, Gay, preached to my son to get out of Telluride. There was no work. No future here for a young person. My son told me to buy land. We had no idea Telluride would become this popular. The popularity seems odd, but since the mine closed anyway, it was a blessing the ski area came in or we'd be a ghost town.

When they laid me off it was about 1995. I thought, what am I going to do with my time? But I've got plenty to do. I have five condos to take care of. And I work in the cemetery during the summertime. I drive the senior citizen van. And a couple of old ladies that live up here, I do all their snow shoveling for them. My wife worried about me when I was in the mines, but I don't think she worried about me as much then as she does now.

Gay and I eloped when I was eighteen years old. My cousin and his wife went with us. We told our parents we were going duck hunting. Oh if you wanted to see one mad woman, you should have seen my wife's mother when the paper came out and there it was in the paper that we'd got married. But it panned out good for us. Fifty-seven years. I'm a good, cheap cook and my wife doesn't mind cleaning up, so we get along good. I hate to clean up.

David Calhoon

Ouray, Colorado

I was born in 1924 and was raised in the Cripple Creek area. Most of those mines in that district were owned by the Carlton family, d.b.a. Golden Cycle Corporation. They owned the trucking company. They owned the railroad that took the ore down to the mill. They owned the mill. They owned the bank. So it was pretty much a company town.

My grandfather homesteaded a ranch out there, and my dad grew up there. I was about four years old when Dad moved the family to Oak Creek, close to Steamboat Springs, so he could work at a coal mine up there. He bought a team of horses and a wagon. He then took that wagon and he made bows out of pipe and put a tarp on it. He put our beds in there. There were four of us boys and my mom and dad. The beds, the cook stove, the six of us, and everything we had went in that wagon. We took off in May.

We traveled with that team and wagon all the way from Oak Creek clear over to a little railroad town called Minturn, where my dad's half brother lived. So we camped there with the wagon, and then my dad put the team to work building trails and stuff for the Forest Service. He worked that team for about a month—actually made pretty good money.

Then we went up over Tennessee Pass into Leadville and on down to Buena Vista where we had some relatives. Camped around there. Then went on into Cañon City. Had some more relatives there. Went to Florence. More relatives there. And from Florence we went about thirty miles up through Phantom Canyon to Victor. It was about the latter part of August. We camped out on what they call Pickle Hill until my dad finally found a house to rent. And then he got everybody in school.

After about two years, Dad bought an old house in Victor. Oh it wasn't much of a house. He built a couple of rooms on over the garage part of it. Us four boys, we slept in this little room down in the basement, in two double beds. We didn't have any indoor plumbing. And Victor had a cold climate. It was 9,700 feet in elevation. It was raw. Never did snow much, but the wind really blew it around. Some places it would be completely bare, and other places there'd be drifts eight foot deep.

Eventually Dad bought another house. He paid $800 for it. We were really "uptown" then, because it had indoor plumbing. But it was only a little better than the other one, because it didn't have a hot water heater and there was no insulation. When the wind would come up from the south, the wallpaper would blow off the walls. But it wasn't too bad. Just cold.

When I was in high school, I worked at the mines during the summer, digging ditches outside. I wasn't so much interested in mining. It just was about the only job you could get. One winter I worked on graveyard shift at the mill. I'd go to work at 11:30 p.m., and get off work at 7:30 a.m. Then I had a paper route. And then I'd go to school.

During the Depression, when there wasn't any other work, those mines in Victor were all running. Some of those fellows that came over here, I remember them in the bar when they got their paychecks. They could drink for days. I never ran into anyone with that same caliber of drinking.

During World War II, they closed down the mines in Victor, because all they were mining was gold—no strategic metals. I went into the service. And after the war, those mines were running full blast again. I went underground then.

Most of my experience I just picked up as I went along. Watched what people were doing, paid attention. All the jobs in the mine have a certain amount of risk. They had problems with bad air—gas. It's real strange. You can't see the gas, and you can't smell it. It was like a doorway. You walk down a drift and the air will be good, and suddenly you can step into bad air and you can't hardly get your breath.

I was on the skip with my helper, bringing guys up and down. One time the mine superintendent rang us down to the ten hundred level. We went down, and he said that there were two guys he couldn't locate, and he was worried about it. He wanted me to go with him to look.

I grabbed a light, and we got back to this certain place. A tramcar was there, and he got hold of it and he said, "This is as far as I could get before." I noticed it was suddenly hard to get your breath. He kept pushing the tramcar, and I was following behind him. Pretty quick, he just flopped. I thought maybe I could get hold of him and drag him back, so I reached down and I just felt weak all over. Something told me I better get out of there and get help. I ran back to the station, got my helper, and we went back there. Where the superintendent went down wasn't more than twenty feet from where the air was good.

My helper said, "I'm fresh. You hold the light, and I'll run in there and get him." He grabbed his nose, and he ran in there, and he got hold of him, and my helper just turned around, and his eyes got big, and he flopped in the middle of the track.

I thought, "Well he's a little closer, maybe I can get him." So I set the light down, and I got in there a ways, and I knew I couldn't do anything with him. You can just feel it. The bad air takes over so quick. So I got out of there, and I went back, and I got on the skip. I rang to the top, to the master mechanic and said, "I need some help down here. I got two guys down in the gas back there."

"Oh, I can't leave," he said. "I gotta stay up here."

I stepped back outside the skip and saw three guys that had changed their clothes to go home. I told them what happened, and they didn't change clothes or nothing. They ran and got their hard hats and their lights, and they came out and I took them down. First thing they did— I never had any experience with this, so I didn't know what to do—they grabbed a pick. And they went on back there and when they hit the gas, they started picking holes in the air lines, to get some air in there.

They got them out. In the meantime, we called the doctor from Victor, and I gave him a ride down to the ten hundred level. They had my helper stretched out on a bench, with popped-out eyes rolling back in his head. But he was alive. The superintendent was dead. He was forty-five years old. A young fellow, really.

The guys that were missing, they'd worked there a long time, and they knew gas. Turns out, they'd run over into another mine there and were prospecting, so they were fine. They worked their way around and climbed up and came out on a different level.

But that was a bad experience for me, because I didn't know what to do. I'd never been in that gas before. I didn't know anything about it. My helper never worked underground after that. He quit.

When I first came over here, to the Idarado, I lived in the boarding-house. It was real good. They had two beds in a room. And they had a commissary, where you could buy your clothes, your boots, your hats, cigarettes, candy, and beer. And the boardinghouse served breakfast. It was "all you can eat," and then they had stuff for your lunches. And then they had supper. At supper you could fix your lunch if you were on night shift, which went to work at 6:30 at night. When you came off the shift at 3:00 in the morning, you could eat breakfast. So they were open pretty well all the time. And meals were good.

I had an old fellow who was my roommate. He was driving drift. I think he was close to sixty years old at the time. His partner was probably sixty-five. And they were both little fellows. They both chewed snuff, and they wouldn't talk much. They each knew what they were going to do, and they'd go in and get their drilled round in, and their round out, every day.

I worked in Leadville for a little while and worked in uranium. Then in the 1950s, I got out of mining. I worked up at the Camp Bird for a while, but up there I just did building work and stuff. I did go underground and put some chutes in there, but pretty much I got out of mining.

I became a county commissioner at one point. And I was always a volunteer fireman. We dug people out of avalanches. Most of them dead. Just digging out bodies. It's unbelievable, the force that those avalanches have. It's terrific. There was a fellow from Durango came up, and he was running one of the state dozers up there, an old D-7. It had just come up from the Durango shop and was freshly painted orange.

He finished plowing out a slide, and the road was open. Some people came by and said they'd noticed that he was sitting in the tractor eating lunch. Anyway, there was another slide come down, and it took that Cat clear across the canyon.

When we got down there, we got two tractors to get hold of the Cat and pull it up. Those arms on the side were bent, but there were no abrasion marks on the new paint. No stuff went by it, no rocks or anything. I think it's because of a force that's in front of those slides. Those slides come so fast, they're pushing air in front of them. All his blood vessels were filled up. I think you're dead before the snow ever hits.

One time a slide came down near the Camp Bird Mine, and they blew the fire whistle. We took off up to where this guy was buried. We had to walk. It was about two miles, and it was slow on account of the snow.

When we got up to where he was supposed to be, we heard this Cat running. So we thought, we'll just wait here until the Cat gets down here. We sat there and all the time there were slides running, just booming all around us. After a while the tractor quit running. So we thought we better go see what's going on.

David Calhoon and his grandson, Aaron, at Greyhound Mine, Ouray County.

There was just a black hole coming up out of the snow. That was the Camp Bird's tractor down under there. See, unbeknownst to us, while we were waiting there for this Cat to come dig that fellow out, this other slide came down. Apparently it went up on the other side, up in the air and then it fell. We figured, when those guys saw that slide coming, they might have jumped behind that tractor. So we dug. That stuff was packed. You had to cut it like blocks of ice. There were trees and stuff in there, but no bodies.

What happened, we realized when they dug them out, was that these boys saw it coming and started running. One guy threw himself in the ditch. He suffocated. The other guy was caught in a running position.

The third one got buried, but he lived. The other two had sent him up to the Camp Bird to get their lunches, so he was on the edge of this slide. It knocked him down and covered him with that really fine misty snow that comes afterward. It settles in. He managed to get out of there and get up to the mine, but it affected him after that. He never was really the same. This was on Valentines Day, 1958. It was a bad deal.

Not too many of the men I know are mining any more. We get together and tell stories, usually at high school reunions. But most of the guys are dead or dying. And those who are still mining are either working in Idaho, Montana, or Nevada.

There's more restrictions on mining now than any other industry in the country. Everybody thinks mining is a bad deal. But nobody wants to give up a car, television set, or telephone. They don't like logging, but they all want log houses. We're a consuming country. We consume, per capita, a lot more than any other country in the world. And everything you use comes from the earth. All your food comes from the earth. All your materials come from the earth.

I, myself, was never really into mining that much. I liked prospecting. I enjoy going into old workings and looking it over, trying to figure out what the old timers were doing. I enjoy that more than the working end of it.

But, of course, if I had a nice streak of gold, I think I could probably work pretty hard at it.

Aubrey "Blizzard" Lillard

Eckert, Colorado

I started mining at the Rico Argentine when I was nineteen years old. It was around 1943. There were about four houses in Rico then that had indoor toilets. The rest had outhouses. Used Montgomery Ward catalogs for toilet paper. Sometimes the snow was as deep as your neck, so you couldn't tear along out there.

They had a lot of Indians working there, more Indians than white guys. The Indians had better houses. They had toilets and everything. There was this guy that needed some Levi's. The shop owner never did have any for the whites. But this guy went in there one day and said, "All my neighbors are complaining."

"What about?"

"Well, my pants. They're so raggedy. I see these Indians all got new Levi's. If you can see your way to it, it would sure be appreciated if you let me buy a pair." He said, "It's not that I want to, but the neighbors are complaining about me looking like this." She finally broke down and sold him a few pair. Then everybody went down there wanting some. They found out she was charging one dollar more to those Indians than she should.

She was going down the road to Cortez, one time, and a bunch of elk run near over the top of that brand new Buick she had. That was really sickening to her, but it tickled a lot of other guys.

Hardly anybody else had a car. Every time you'd buy a new car, the dang mine would shut down, and you couldn't make your payments. So people would drive their cars over that steep place and call it an accident. I thought about it, but I loved my car. And they'd have caught me just as sure as thunder.

I went to work at Telluride Mines. We wore army wool coats and rubber pants, bib overalls. They didn't have insulated underclothes then. If you got a leak in your rubber boots—Oh, the old toes about came off!

You'd get soaking wet in there. And in that time, we didn't have a place to wash up. We had to go home like that. There was warm air in the room where they dumped the ore. So we'd sit in there to warm until the frost melted off our clothes.

It was miserable, there was no doubt about it. And the hardest part was the noise. Those machines put out a racket. Most people have been around outside when jackhammers were going. Try going into a little eight-by-ten-foot place and turn two of those machines on. After I worked all those years, they got us ear plugs. Told me I had to wear them. But that was like shutting the gate after the horses got out.

We were working in some bad country one time, driving a drift. My partners on the other shift had gone in there and slabbed off about forty feet. The next morning, I went in with my shift partner, Ed Baker.

Aubrey with baby Sharon in 1946. Courtesy of Sharon Lillard Albin.

I was looking at that place, and it really looked bad. The walls were popping. I grabbed my bar and ran. I yelled to Ed, "Get out! Get out!" I could just feel it coming down. Well he sort of started running, and I said, "Get out of my way, man! Go!" He wouldn't do it. He just kept trotting along. I started to hit him with the bar. I wanted to run right over him.

When that fifty-ton slab came down, it just nicked me and blew me out of there like a champagne cork.

Busted my hat. I was so sore. Black and blue all over.

Ed was the father-in-law of one of the boys—our partners on the other shift—that had not barred down that area correctly. Ed said, "I think I'll go home and kill my daughter's husband. He almost killed us." He asked me how come I knew to get out of there.

I said, "I've had quite a few years experience in there and I could just feel it coming. And next time I say run, you better move!" I sat down and I said, "Can I have a cigarette, to settle my nerves?"

He said, "Yeah me too. But I'd rather have a double shot of whiskey."

Then we had to get back up and go mine. I was a little nervous for a day or two.

Aubrey in 1946, Rico, Colorado. Courtesy of Sharon Lillard Albin.

∞

Another time the same guys on the other shift said they had mucked out and barred down. All we had to do was set up and drill. So Albert Moureaux was drilling down on the right side, and I was drilling on the left side. Then he just disappeared. I looked down and Albert's laying flat out on his back. A rock, about eighty pounds, had fell on him. Broke his hat. It looked like you could see down into his brain. He was breathing, but he was swallowing his tongue. My old hand was dirty and greasy. At least I did take my gloves off before I pulled his tongue out. I turned him over and drug him back from the face.

Albert was a 200 pound, great big ol' guy. I couldn't load him myself. There was nothing to do but hop on the motor and get some help. So I ran down there about half a mile, got the stretcher and three other guys. We started back, went around a corner and there stood Albert. He said, "What the hell happened?" He sure looked a mess.

I said, "Lay down on this."

He said, "I ain't gonna lay down." We couldn't make him. He figured he'd die if he did. That's what he told me later. It turned out those guys had neglected to bar down like they were supposed to.

Albert was off for about a month. He was quite a guy. He lived to be a dang good ninety years old. Buried in Telluride. I was his pallbearer.

Every time one of my partners quit, there was a bunch of guys wanted to work with me. I usually picked somebody who didn't know anything, because they don't argue with you. When somebody knows something, you've got a problem.

The guy that gave me the nickname "Blizzard" was old Albert Fish, I think. We were working way back in the old Black Bear Mine in this thirty-three degree water. Cold! You'd go back in there, and if you didn't make it in six hours, you'd freeze. All the guys behind us were stoping. They'd blast, and it was just smokier than all get out. You couldn't see anything. And when we'd get done, we'd hop on that five-ton battery motor and just open it up. Haul out of there like a blizzard.

One time we were heading out of there, and we ran under a chute and—Oh, no!—there was a three-car man trip, with men on it, just sitting there. We must have been doing fifteen miles an hour when we hit. Knocked the eight wheels out from under it. Guys went every direction.

They weren't supposed to be there. Two guys were deliberately holding them up, trying to play a joke on me, I guess. But that wasn't much of a joke. One old miner said he was wandering around there, and if he hadn't just stepped off to one side when I hit, the impact would have cut him square in two. The guys that did this didn't work there much longer.

I don't really miss mining. Too many people died. There was that young kid who was real popular. A mechanic helper. He fell in an ore pass. They could see him down there, about 350 feet. The bosses wouldn't let anybody go down and get him. They were afraid to cave in the walls and kill somebody else. They watched him for twelve hours. Said he never moved. Eventually they just pulled him on out from the bottom. He was the lone kid in his family. And they were wealthy. He had no business being in there, but I guess that was the way he wanted to go out.

You get used to it, the accidents and people getting killed. But the first ones just about destroyed me. Worrying about their kids and their wives. It was a catastrophe when they had a bunch of kids.

One of the worst accidents, I remember, was that young guy who had a wife and two or three kids. He set up drilling his first hole and a big old slab came down and squashed him. He didn't know the country. You have to know that country when you start to mine it. It talks to you, and you better listen.

The guy who took his job, that was a sad situation too. He was about twenty-three years old. Broke his back. We were packing him out. We just had this little old narrow skip, an elevator type deal. We had to stand on each side and hold him, standing up. His leg was about cut plumb off, I mean just hanging by the muscles and bones. He didn't say a word about that. He just said, "Oh my back! My God, my back!" Of course standing him up like that is probably what did it. Cut those nerves in his spine. He didn't lose his leg, but he was paralyzed.

The closest I came to dying was when I was running a mucking machine. We were starting a raise. I told my partner, "You set up there, way over on that side." I knew it was bad. I was getting the dippers rolling, pulling the car. And all at once, that old mucker started to flip forward, and I released the dipper. Big ol' slab came down and tore the whole bottom of that bucket out. I would have been right under it. I said, "Why didn't you holler?"

He said, "I didn't have time! I just blinked my eyes, and it was there!"

They call mucking machines "widow makers." A lot of guys hit themselves in the head with the buckets. When I was first learning how, I really thought I was good. Boy, I could really make that old mucker get up. Then I got too much slack in the chain one time. When it buckled, it turned over on me. Broke my tailbone. I couldn't sit down or even lie on my back for about a month. I never did go to the doctor. Eventually I just got back up on the mucker and went to running it right. That's all you have to do—just run it right.

Everybody who works in the mines has close calls. It's easier when you don't ever think about them. That big slab that came down that time about scared me half to death. I've drilled under a lot of bad country, and I sure was nervous. I tried to get where I could always have an opening. You keep looking up. And sometimes you hop, because you know it's going to get you.

I mined in Telluride for twenty-two years. Then I left to be a mine boss down in New Mexico, near Gallup. Uranium mining. I "mine bossed" with a bunch of Indians, and that was something.

One day my kids came down and said, "Dad, you better get up there quick. A man's killing his woman."

I said, "Aw, get outta here." I was reading.

"No, we're not kidding."

About that time—those Indians never knock, just walk in—this Indian woman came in and said, "Man killing woman pretty quick."

So I jumped up, got my hat and shirt. I hopped in the jeep and ran up to where they had Indian housing. They were one-room deals. Probably three families living there.

Boy, he was really tromping the heck out of her. I went over and yanked him off. He dropped back. I said, "I wouldn't advise that."

He said, "What are you gonna do?"

I said, "Fetch me your wine."

He said, "I drank it."

So I brought out a radio and said, "I'm gonna call the Indian police."

"Oh, now wait a minute," he said and went and got his wine. I broke it over the jeep. Boy, he didn't like that at all.

I said, "One more thing. I'll let you know I'm calling the Indian police."

So he made himself disappear. After about three or four days, he came in and went back to work like he had never left. I heard a commotion. I looked up and saw two great big guys beating the living heck out of that guy. They were using clubs. I hollered, but I had to run and jump in the jeep and go up there. I said, "What the hell's going on up here, beating up my man?"

"Oh," one of them said, "we're the Indian police. There's very few of us. So whenever we do get around here, we try to make a lasting impression." The guy got thirty days for beating that woman. After thirty days, he went back to work. Never said a word. They'd just come and go like that.

The doctor down there told me I had cancer, so I left. I moved up to Eckert to retire. The doctor up here says it's not cancer. It's silicosis. He gave me a respirator.

The other day the doctor said, "I'm gonna have to cut your leg off." Then he left the room. After about thirty minutes, he came back and said, "Well, I'm not going to have to cut your leg off. But I'm going to cut your foot off."

I said, "Well at least then I can probably still reach the clutch on my pickup."

I didn't think I had the energy to get out of bed today. But I got up. Somebody's got to mow the lawn around here. I never thought I'd live this long. But I also never thought I'd wind up like this. If I'd have known, I would have quit the mines long ago.

Dr. Edward Merritt

Cortez, Colorado

I hadn't imagined myself working for a mining company. But I was twenty-seven, it was 1948, and I needed that additional income to survive. My wife and I would go up to Rico every Thursday afternoon to see the patients. That was my day off. They had some 225 miners up there when I started. I would take care of them at $1.50 for single men, $2.50 for a family. Some of the Navajo interpreted "family" as anyone related to them. They would bring in their entire families. Aunts, uncles, and grandchildren.

I kept a little clinic in one of the old mine buildings. I had a used x-ray machine, and I got all the drug samples I could. I had a practical nurse up there. Her husband was a miner. She would telephone me and say so-and-so is sick, and this is what's wrong with him. I would tell her what to do. She used to give shots of penicillin. Strangely enough, in all the time she worked up there for me, she never had anyone get a bad reaction. Down in Dolores a nurse would give a shot and some of the patients would be on the floor before the needle was out, because they were allergic to it.

The nurse's husband and his brothers leased a mine. When the Bear brothers finally ran into some productive ore, the mine manager decided to not renew their lease. The day after the brothers were told they couldn't renew the lease, the stope where they found the ore "accidentally" collapsed in an explosion. Nobody was able to find that vein again.

∞

I had one patient whose name was Bochinclony, which means Many Horses. Every Sunday morning he would come by my house and wake me up to pay me five dollars on his bill. He was probably going home from his shift. In those days you accepted barter—chickens, geese, and eggs, whatever they could pay you.

A lot of the miners working in Rico were Navajos. The union tried to come in and unionize them. Got them all drunk one night. Well most of them had been in the service, as code talkers. The Germans couldn't understand or de-code Navajo, because it was such an unknown language. So the Navajos were a big asset in the war. When they got drunk that time, they wound up bringing out their sabers and swords and mementos like that, which they had brought back from the war. They had a real furor that night against the people that were trying to unionize and went on a rampage. But it eventually simmered down.

I had to make an emergency run up there to the mine in Rico when one young miner was killed. He went into the mine up on the hill. One of the reasons why those miners used candles in the mine was because if the candles went out, it was from lack of oxygen, and they knew there was mine gas around and that they'd better get out. But when this miner's candle went out, he just went and got an electric one. He went back in there and the bad air killed him.

Two other miners got caught in a mine gas deal up there. One of them passed out. The other miner cut an air hose, so air would blow in his partner's face while the other miner got out and went for help. He then went back to get his friend. They were both brought to an area in the mine where there wasn't any danger, and I went in to them. That was really scary. I had a headache from the time I went in until the time I came out. I didn't know whether it was a noxious headache from the explosives, or the gas, or the altitude, or just nervous tension.

We put the two miners in my car, and we headed to the hospital in Cortez. Up there if you waited for an ambulance, your patient might die, because it's about fifty-five miles to Cortez. I had Wilson Brumley ride with me. He's a rancher out of Dolores. Halfway to the hospital, I ran out of supplemental oxygen for the miners. That was the last time Wilson ever rode with me. I guess it had something to do with the rate of speed I drove, but both miners were saved.

I did not relish going into the mines. One time I had to go into the Telluride Mines. It goes underground all the way through the mountains and comes out on the other side. You'd go from a twelve-man skip, which is like a big elevator, to a six-man skip, into a four-man skip, into a two-man skip. Then you get in the one-man skip. You're in those skips, and some of them follow the curve of the wall and, well, it gets fairly creepy. I had a headache the entire time.

When you get out of that, you wind up hundreds of feet under the ground. You turn around, and the skip is gone. You see a big black hole, which if you fall in, you don't know where it's going to end. You've got to back up and get your feet on the ladder to get down the hole to the next level. You're climbing down with your hands on the rails, and if you joggle your light or its cord going to the battery at your waist, the light might wind up pointing at the ceiling instead of down, where you want it. I could never understand how the miners surveyed underground as well as as they did.

I suppose the most common mining accidents were foreign fragments in the eyes. The miners had to clean out the air compressors, and when they took the cap off, the air would blow dust out and impregnate their eyes with it.

Also rocks fell on them, and there were a lot of hand injuries. If their motor belt was a little frozen, the miners would grab the belt with their hand and give it a jerk, to break it loose. The freed belt would take off and sometimes catch their hand and just gruel it right down to the bone. In those days we would clean the wound and trim the edges, because the rubber was impregnated in it, and run sutures through it.

But the most common problem was arthritis, because of the dampness in the mines. Also pneumonia. There's decreased oxygenation above

6,000 feet. When you get up to 10,000 feet, the air is getting pretty thin. Decreased oxygen in the blood retards healing.

Yet the people of Rico were dedicated to living up there. You couldn't get them out of Rico. A lot of them died up there. Orval Jahnke, one of the last mine managers, came home one day and was carrying some groceries into the house, when he collapsed of a heart attack. Orval's wife called Dr. Houston, who had trained with me and just bought a house in Rico. He had to go up into the attic to find his bag, but he got to Orval and was able to keep him alive with CPR and adrenaline medication. Orval was shocked several times during the fifty-mile ambulance ride to the hospital. When I met up with them at the hospital, Dr. Houston stated that Orval was "gone."

I said, "No. We've got to save him." I continued with the treatment, and Orval came around. Then I worried he'd be a vegetable, because he'd been without oxygen for so long. But Orval woke up and was ambulatory and lived ten more years.

I enjoyed medicine. When I was eight years old, I had a ruptured appendix. I wound up having three operations. I got into medicine for the benefit of making people well.

One time I had to go part of the way on horseback to get to a patient who had pneumonia. He lived outside of Dolores. They met me on the main road with a horse, because the way to the house was too muddy. I always wore a suit and tie. But I took my bag, got on the horse, rode to the house, and took care of the patient. Incidentally, he was one of the two men who, some time later, froze to death carrying money to the general store.

The general store was where miners cashed their checks. The two men were driving a new Dodge pickup. It was winter, and I guess the truck was so new it hadn't yet been put to the test. They picked up a couple of Indians along the way and, at one point, the truck just gave out. They couldn't get it started again. The Indians took refuge in a shelter, but the two miners figured they had better get the money to the store, so they set out in the thigh-high snow. After a while one of them wound up having to carry the other. Eventually they both gave out. They were discovered the next day at Scotch Creek by the men who plowed the road.

It snowed a lot in those days. One time old Peabody had a heart attack. He had been a miner and was now retired and living in Pleasant View. I told his family, "Well if he's having a heart attack, take him right to the hospital." But they insisted I come out. It was cold, snowy, and windblown. That was a hell of a storm that night. I went out there and had to really maintain to get through to him. I gave him a shot, and we put him in my car and got him to the hospital.

I was the only doctor I know of that made house calls in the area. I delivered babies in peoples' homes. No matter what the situation was, I always made sure to wear my suit and tie.

There was this old miner who liked to tease me because, one time, he called me at 2:00 in the morning, and I beat him to the hospital and still showed up wearing my suit and tie. I always thought if you were a doctor, you should look like one. I don't go for the sloppy way a lot of young doctors dress now.

Myron Jones

Rico, Colorado

I was born in 1924, and we moved to Rico so my dad could mine. Trouble was his life. They didn't have cars or trucks. They only used horses and wagons. The dirt road to town had two big ruts that your wheels went in. One time my dad met some other guy coming the other way in the ruts. The other guy said, "Pull over."

My dad said, "You pull over." So they got out the whips and started whipping each other's horse teams. It was such a mess of whips and horses, it wasn't really clear who gave in.

In Rico we got around on donkeys, horses, snowshoes, and skis. Skis were different in those days. They were just a couple of boards with the ends turned up. Probably weighed about three times what skis weigh now. However you got up the hill to the mine, you packed your skis up with you and went down on them. You had to enjoy skiing a lot then.

When my dad and his partner were driving a tunnel up here in his mine in Rico, he took me with him one night to show me what he did, I guess. I remember going in with him. The air drills. I could hardly stand the noise. It just banged on my ears. But I stuck it out.

I did not think mining was something I wanted to do, but there was nothing else to do around here. So I started mining when I was in high school. I worked on jobs outside the mine, but sometimes I'd go in.

I started out as a mucker. When you were new to a mine, it didn't matter how good you were, or if you'd worked for years in a different mine, you started out as a mucker. Then they'd figure out what you could do. But my dad had influence. They moved me up right away. Then I worked at everything there was to do.

What bothered me most was the noise. You talk with your hands. It took a while to catch on to. I suppose you could call it a kind of a language.

My main job was what they call pulling pillars; the pillars of rock that hold up the walls down there. They get some idiot like me to come in and shoot the pillars out after they're through. You go in and blast out whatever is holding it up. Stays up for a while. Sometimes it would come down; but, hopefully, not while I was there. Of course it's not too safe a job, but it's more money.

I had a few close calls with my partner, Tommy. Tommy and I were great friends. Sometimes the rocks would start falling. We'd run and huddle under some place where they weren't falling, until it quit.

I suppose there were some safety regulations, but we kind of made it up as we went along. Didn't really worry about it. People get hurt now with safety regulations. The key is to take care of yourself. Watch what you do. The way we built up timbers, to secure areas, was better than the way some people build houses. They were beautiful.

I did have a few of them hard hats broke. There was one slab I worked on a lot. Every time I went by, I tried to get it down. One time I went by, and it came down by itself. I swear it waited until I was going by. My ears rang for a while. The slab also gave me a scratch on my head that the doctor stitched up for me. He was in town that day. I gave him a gold specimen instead of paying cash.

There was a lot of high-grading, which is when you collect a pocket of gold or something, and you sneak it out and take it home. Everybody high-graded. Kept a specimen stuffed in your boot or lunch pail.

When Tommy and I were shooting out pillars, we were very careful to keep it so unsafe that no inspectors would come around. We found a streak of gold there one time and took a few specimens out of it. But usually you didn't have time for that.

After a while I kind of got antsy and took off for Alaska. Then my dad got a lease at a mine here, so I came back to help him.

I never was too enthusiastic about company mining. It was just a job. Kind of dark and dirty. But when we were working on our own, I liked mining. I liked the independence. I got kind of fascinated with it.

I had some partners I didn't get along with. I like deliberate people. When people bang and throw things around, it makes me nervous. I had a partner named Buck. Buck was what you'd call a "mountain man." If he wanted to get an elk, why he'd just sneak up and shoot it in its sleep. No chase there. One time, he shot an elk, and he came back for it the next day. A bear had drug it off and covered it up. Buck uncovered it and took it anyway.

Years later, Buck was working in some mine. His partner went down the ladder one morning, and he collapsed. He'd gotten into some bad air, or gas. So Buck went down and dragged him up and laid him down. Then Buck laid down and died. The gas, plus the exertion of carrying out his partner, killed him. Strange thing was, the other guy lived.

I got caved in on one time. I was alone, so I had to dig my way out, up where there was air. I didn't even consider that I was dying until I was out on top of the pile. I sure thought about it afterward.

I went to work up at a mine on Lightner Creek. It was leased to the VCA [Vanadium Corporation of America]. They sent me up there. I messed around by myself for a while. When my brother graduated from Fort Lewis College, he came up and stayed with me a while at the Lightner Creek Mine.

We worked together up there and used to go to Durango and party all night, then come back up to the mine. The road crossed Lightner Creek—just went straight through the water. One morning we got stuck right in the middle of the creek. There was water pushing against my door. I opened the door anyway, and my brother opened his door, and the creek went right through the car. So we got out the jack and then jacked the car all the way across the creek, six inches at a time.

Some time later, the VCA sent some supplies up there. Same thing happened. They got stuck right in the middle of the creek. So we had to carry our supplies all the way up to the mine.

My brother and I left the mine at Lightner Creek and moved back to Rico, to do some assessment work—looking at mines. We went down into the Iron Clad Mine and found a fissure of gold. We just climbed up in there, and there it was. We sold it to the Denver Mint. We were paid several thousand dollars. I bought a new car and my brother paid for his college.

All my folks mined. I think my mom thought I should be a doctor or a lawyer. She was an old Irish lady. She had her mind made up about a hundred years ago and wouldn't change it. She didn't put up with much nonsense. She was real proud of my younger brother, for his education. I never did go to college. Never had the time. I just always had something else to do. He wound up becoming one of the "big wheels" at the Climax Mine.

I was the deputy sheriff for a little while. I kind of enjoyed it. There was trouble all the time, but nothing you couldn't negotiate. I'm six-foot-four-inches tall. It was always better to talk your way out of it, as opposed to going seventy-five miles to the jail. The county seat was in Dove Creek. If you took anybody in, you had to take them all the way to Dove Creek. Mostly I didn't consider it my job to arrest people. I was there to help them out, and just my presence would help calm things down sometimes.

I stopped mining when my health got me, in about 1996. My kidneys went bad, so I have to do dialysis. I'm old, so they're not so enthusiastic about a transplant.

I'm real sorry that they don't mine here anymore. I suppose it's just cheaper somewhere else. It's regulated to death here. Safety regulations, clean-up regulations. Here in Rico, they put in millions of dollars just because two real estate people didn't like the looks of that mine over there.

They pushed that mine up against the hill and tried to grow grass on it. It looks like hell to me, but they like it.

Al Maes

Interviewed in Silverton, Colorado
Lives in Arizona

Growing up in Silverton, my best friend was a little guy named Billy Rhoades. We were like brothers. We did our first mining job, outside the mine, when we were very young. It was during the early part of World War II, when miners were in short supply. Some of the real miners got deferments because they were needed to mine the precious metals. The others were either drafted or volunteered for the service.

Since all our natural resources were being used up for the war effort, the government was putting out loans for people who were interested in trying to develop new resources. So this guy, Joe, who used to run a pharmacy here, took out a loan. But he didn't have any miners, because the miners were already engaged in mining or had gone to war. So he hired Billy and me as young kids, fifteen and sixteen years old, respectively.

Joe sent us to Chattanooga to get an old dump truck. It was a rickety old thing. Of course I'd never driven a dump truck and neither had Billy. The first few miles I drove it, we were grinding the gears, but we finally got it to the Mayflower Mill.

We gathered up the ore that Joe wanted to ship by rail to Durango, to have it analyzed. We loaded it up, and we were coming down Main Street in Silverton. Billy had these little short legs, so when he kicked back to put his feet on the dash, he inadvertently put his foot on the handle that engaged the dump. When he did that, unbeknownst to us, we dumped a big pile of ore right in front of City Hall.

We just continued traveling down Main Street. The town marshall caught up to us. He said if we didn't get it cleaned up, we were going to reform school. So we went out there with shovels and mucked it up. That was our first experience with mining.

Billy and I always seemed to be getting into some kind of trouble. One time we were having a party, and Billy came up and said, "Want to go for a ride?"

I said, "What do you have? A bicycle or what?"

He held up some keys and said, "I have a car. This guy, Weaver, loaned me his car." So we jumped in the car and headed up to the Mayflower Mill. Just joy riding, you know. It was late at night, and we went around a curve real fast. Billy was so short, he could barely see over the windshield. And here comes a bus from the mine, with a crew that was coming off a shift. He side-swiped them and knocked them off the road.

All the miners got out. My dad was one of them. He chewed me out. They gave us orders not to leave and not to move the bus. Then they took off and walked the rest of the way to town. The car was almost totaled, so we went inside the bus. Billy said, "I wished I had a cigarette."

I said, "Well let's go get some."

He said, "I'm not walking all that distance." So we managed to drive that bus to get some cigarettes, then went back and parked it exactly like it was. And we went to bed in the bus. Next morning the highway patrolman came up and got us, and he put us in the backseat. We were going all over: to the courthouse, the city hall, up to our house. All the kids would see us in the back of the patrol car, and we'd be waving at them. They thought we were heroes. There wasn't too much to do around here for excitement.

By the time I hit junior high, the war had broken out in earnest. I decided to go into the Navy. Mr. Richard, the school superintendent, was so irate that I quit school, he told Billy, "There goes a lifetime private." I wound up attaining the noncommissioned officer highest rank possible—chief master sergeant; but the superintendent got over being mad at me even before I left. The day that I got on the bus, he let the whole school out to say goodbye to me.

It was 1944. I had just turned seventeen. I'd never been away from a small town like Silverton. We were kind of a protected group up here. One time we had a basketball tournament in Denver. My cousin was there, and we saw these streetcars. And my cousin says, "Look at them big yeller trains!" Everything was a new experience to me, a big challenge.

When I got out of the Navy after four years, I came back and went to work at the Mayflower Mine. I almost got killed up there. They had me working in a tunnel. We were hauling ore, from the ore pockets, and I was the motor man. The motor was run by electricity, and it had a trolley pole. You'd come in, and the trolley pole was behind you. The tunnel was so narrow, you only had a certain spot where you could raise that trolley pole and put it behind you to go out.

I was standing in the locomotive. I held the trolley pole and went to turn it around when the brake dislodged. The cars started moving. It all happened so fast. The trolley pole lodged against the rock wall and caught me on the chin. Of course the cars were still moving; and the trolley pole lifted me up by the neck, threw me down between the first and second car. Some way or another I was able to stand up and get against the wall.

These cars were loaded with so much jagged rocks and stuff. Luckily, I was real thin, and I exhaled as the next car passed; but the corner of the car caught and gripped my chest. It was a good thing I was up against the timber instead of the rock wall. The timber was wet so it spun me completely around; and when the second car came, I fell right between the two cars, onto the coupling that joined the two cars. The train went down until it leveled out and when it stopped, I was still there on the coupling.

They took me to the hospital. I had a lot of broken tissue, but I didn't have any broken bones, so I said to myself, this is enough of this business for me.

I quit that mine and went over to Telluride. Billy was over there, working in a different mine. He'd been in and out of the Navy too. I ran into him, and we went over to the boardinghouse. We ate dinner, and he said, "Why don't you stay here. I'll get you a job."

I said, "No, I'm just cruising around looking for outside work." I wasn't too keen on going underground again. But eventually I went to work over at a mine called the Silver Bell, in Ophir. From there I went to the Idarado Mine. Later on a friend induced me into going up to the Camp Bird Mine, southwest of Ouray, way up in the canyon. I almost got myself killed there too.

I was what they call a timber man. Dick Gray, the boss, sent me up to repair some lining boards in a manway, which is a ladder where the people would go to their jobs in a stope. They'd go through a little hole in the wall to where they were mining on either side of the manway.

I was up there repairing a lining board with a timber hatchet. Above me was another tunnel. They were transferring ore from that tunnel to a big ore pocket. They were supposed to close the trapdoors to the manway, but they didn't do it.

A young Al Maes in 1945.

I had one arm around the ladder as I drove the last spike in the lining board. When I finished, I got ready to walk back down and here comes a whole bunch of rocks. I snuggled up underneath the lining board. There was very little room between the lining board and the open manway and these rocks were just zinging by me.

I knew what was happening. I knew that they'd forgot to close the trapdoor. I also knew the sequence for how long it would take them to fill their train. So I figured, well, if I could just stay here that long, when they're finished, I could go on down to the lower tunnel, the main tunnel.

I couldn't yell. Nobody was there. It was pitch black! The only light I had was my light, and all of a sudden, a sharp rock hit my hard hat. It took my hard hat and cut the cord on my lamp, and it went straight down. So there I was, no hard hat, nothing.

The falling ore finally stopped. I thought, here's my chance to get out of here. I felt around, found the ladder, and I scurried down to the next tunnel. It was pitch black down there. At about this time, here comes the main haulage crew coming down the main level. And they were going full bull with a full load of ore.

Of course they couldn't see me. No light, nothing, too loud to yell. So I hurried and found a place in the rock wall where I could feel that I was safe. But then I was worried about the huge loads that they put in these five-ton cars, hanging over the edges. They load as much as they can, because they're on contract, and they get so much money for how much they haul. I was just lucky that one of the guys turned, as they were going by, and his light flashed on me. He signaled to the motor man to stop.

The motor man screeched the brakes, and they came back and got me. They took me out to the portal and took me down to the hospital in Ouray. They examined me, but I wasn't injured, just badly shaken up by the close call.

After all those accidents, though, I kind of felt like I might be accident prone. But it turned out that was nothing unusual. I mean there was a guy there that I mined with at the Camp Bird. His nickname was "Bumps." He was a great, big, husky guy. They called him Bumps because he fell down a manway, and he lived. He got bumped coming down the manway, hitting those timbers, and it broke him up, but he lived through it. Things like that happen when you mine long enough. My father once told me, "Son, mining is a wonderful, honest way to make a living, but in the end, you'll either get silicosis, or you'll get maimed, or you'll get killed."

My father, Ambrose Maes, was a mining superintendent in Salida in the 1920s. He married my mother when she was sixteen. Pretty soon, my sister, my brother, and myself came along. I was born in 1926. The Depression was threatening. My father could see the handwriting on the wall, so he purchased a ranch down in the San Luis Valley of Colorado. He wanted us to have a place to stay, to ride it out. He remained in Salida until they closed out the mines.

Just as soon as the Depression was over, he moved us to Silverton. I was only about six years old. That was about the time Billy and his family came here too. My father went to work at the Shenandoah Mine.

He worked there for some twenty years, riding back and forth on that suspended tramway. It's got these cables that go clear across that canyon, and little tram buckets that they used to haul the ore from the mine to the mill. But the miners also used to ride them to get to and from work. In the wintertime they had blankets to cover themselves up.

When I got old enough, my Dad would get me summer jobs up there. One summer I was working up there, and I was living at the boarding house. When I didn't have anything to do, I'd go into the shift bosses' office and talk to them. They were all friends of my father. I was sitting there in the office, and the red light came on. That was the signal that there'd been a mining accident. As if by a coincidence, my brother, Leo, happened to be the main motor man that night. When that red light came on that meant they were going to be hauling somebody out. I turned to the shift boss, and I said, "I wonder who the poor soul was that got hurt tonight?"

When my brother brought the haulage train out, they stopped at the office. I walked out and there's my Dad on a stretcher. He had been drilling a stope, and the staging boards had cut loose. He fell a long way down.

They put him on the stub tram, which brought him down to the main tramway. They loaded him on the first bucket. My brother was behind him in the second bucket, and I was in the third bucket. So we headed down to the mill. When we got to the end tram house, there was an ambulance waiting for my dad. They off-loaded him. Then they brought my brother in.

The tram was usually shut down at different intervals at night, when they were finished hauling ore. I guess they got caught up in the heat of the moment, and when the tram operator brought my brother in, he closed it down. They all took off and left me out there stranded over that real high span. I was out there by myself in the dark. It finally dawned on somebody that I was out there. So eventually they came back and brought me in.

They took me to town, and I went to the hospital. They had my dad's mining clothes wrapped in a bundle. My older brother stayed there with him. My other brother, Rich, was playing basketball at the gym. He was a great basketball player—All State. So I walked in there as quietly as possible, to tell the coach that my father had been injured. I said that we didn't know how badly hurt he was, and not to scare my brother. I said, "Just wait until after the game is over and tell him to go straight home."

From there I went home with my dad's clothes. I thought Mom was asleep. I walked in the house very gingerly, but she was standing at the foot of the stairs. She saw my dad's clothes and just panicked. She said, "Your dad's been killed!"

I said, "No, he hasn't, Mom. He's been injured, but it's not life threatening." I didn't even know that. I was just trying to pacify her. It turned out he had some broken ribs and stuff like that. You know, like my dad said, "If you work in these mines long enough, sooner or later you'll get silicosis, or you'll get maimed, or you'll get killed."

After my accidents and because of what happened to my dad all those years ago, I had enough foresight to know what was ahead of me if I stayed in the mines. I quit mining for good, in 1950, and enlisted in the Air Force.

I only went back underground in 1973 as a a mechanical superintendent with the Air Force. I was assigned to NORAD, North American Air Defense Command. It's not too highly publicized.

I retired with over thirty years military service in the grade of chief master sergeant, U.S. Air Force, June 19, 1979.

Billy stuck to hardrock mining. In fact, he spent forty-five years of his life mining. He was one of the best miners around—a mining superintendent and all that. He recently died of cancer.

Clarence and Marie Shilling

Durango, Colorado

CLARENCE: My dad was a mine boss over in Creede, Colorado. I went in there with him when I was about eleven years old. It was around 1941. Back in those days, there were no laws against kids going in the mines. I'd go with him to the working places.

We moved from Creede to Silverton when I was fifteen. They had about seventeen or eighteen mines working up in Silverton then. There were a lot of miners; and there were always some of those guys off work, down in the bars, drinking and fighting.

In the late 1940s I went to work at the Idarado, where my dad was one of the shift bosses. One day he came back on one of those electric motors that ran on tracks in the mine. I happened to be there, at lunchtime. We talked a while, and then he said he was going to go take some supplies to the miners back in there. I threw the railroad switch for the track, and he went back. After about ten or fifteen minutes, a buddy came running down and said they had a wreck up there, and that my dad was in the wreck.

What happened was there was a train on the track coming out with ore, and my dad was on the same track going in. And those guys didn't have any lights on the train, which was against the law. So my dad didn't see the train coming out, and they just hit head on.

I ran back there. A car had jumped the coupling, and it mashed him all down the lower part of his body. We got him out of there, but he died the next day. That was July 25, 1950.

I was just a kid when my dad was killed. Nineteen years old. I went back to work two or three days later. I was kind of scared then. I was working with some old man. We were setting bombs on rocks to roll them; and, I don't know, he was kind of nutty. The way he was doing it, one of the rocks fell out and hit right where one of the dynamite bombs had been. He thought that was funnier than heck, but it scared me, because of what happened to my dad. I quit that day and went into the Navy.

I thought I would never work in the mines again, but then when I got out of the service, my mother and my family were still living in Silverton. Mining was just about the only work you could get there, so I finally went back to mining.

I worked in all of the mines around here, at some time or another: the Idarado, Standard Metals [also known as the Sunnyside Mine], Shenandoah, Camp Bird. We drove the tunnel in from the surface, at Standard Metals. There was an old tunnel that went in for about a mile, and we had to make it a lot bigger. I was one of the first people who worked up there.

I got hurt at Standard Metals one time. I was driving a raise, which is like a tunnel that goes up. I had a timber set up in there, and I was drilling. We wedged in a post-type thing. It broke, and I fell about fifteen feet. I broke my leg and my heel. That laid me up for quite a while.

I also mashed my hand, so it looked like a boxing glove, running a mucking machine. It was a different type of mucking machine than what I'd been running. It was bigger. I reached around to do something, and the bucket rolled back and caught my hand. "Pork" Wilson was working with me that day. He was my partner. He wanted to see my hand, but I wouldn't let him, it hurt so bad. I just ran down the track. But it healed up.

Pork's real name is Harmon. He used to be kind of heavy, so they called him Pork. He and Bill Rhoades ran around together. Billy was a little guy, so they used to call them Pork and Beans.

I knew Bill Rhoades ever since I came to Silverton in 1946. I used to live in the apartment they had by their house. When he was young he was always into something. Ornery! He wanted to fight all the time with somebody. But after he got married, he quieted down a lot. He was really all heart. Everybody liked him.

We went to the Hardrockers Holidays in around 1998. When it was over, three different people came by and told me that Billy Rhoades was dying, and that he'd asked about me. So we went up and saw him before he died.

MARIE: My first husband, Steve Davidovich, had also been a miner. He was working at the Idarado when he was disabled, back in 1963. He

became real sick. He had silicosis, got TB, got everything. But he didn't take care of himself at all. He smoked, and he drank. He was also twenty years older than me.

We had to move to Grand Junction for medical attention. In Grand Junction we found a doctor that used to be in Telluride. He told my husband, "It wasn't so much the mines as it was the rough life you lived. There's worse air in them bars than there ever was in a mine." My husband didn't like that too much.

CLARENCE: Well, back in the old days, you didn't have TVs, or anything like now. Ninety percent of the miners would go out and get drunk. There were very few of them that didn't.

MARIE: He died in 1973. Then my second-to-oldest son, Steve, died in the mine a year later. It's going to be twenty-eight years in July 2002 that he was killed. He was only twenty-six years old.

CLARENCE: You know how the ore pours out of the chutes in the mine. Steve was just a young boy, and I don't know if he really knew what he was doing. He was poking up in there to loosen the rocks so they'd come out of there. He was using a solid piece of steel, like a drill steel, which was against the rules. Somehow the bar got up there, and evidently a big rock hit it, and the bar caught him underneath his chin. They had to get a torch to cut him loose, but he died.

MARIE: He had started there in September, and he was killed the next year. He thought it was a great job, and he made good money. He was getting ready to get married.

He was kind of inexperienced in the mines, and they didn't do the right training. That's a fault of the company but suing them won't bring Steve back. He'd be fifty-four years old, if he had lived.

CLARENCE: When I saw him last, he was working day shift, and I was working night shift. Marie and I weren't yet married. The bus came up the street. He was sitting in there in the window, and he waved at me. About a couple of hours later, his older brother came to my apartment. I couldn't figure out what was the matter with him. He was in terrible shape. In shock, I guess. He told me what happened.

Of course I went to Grand Junction to see Marie. That's where she was living and working. Lee Lopez brought the body to Grand Junction in the back of his pickup truck. The Lopez's daughter had been engaged to Steve.

MARIE: Clarence and I were married the next year, in 1975. I won Colorado Woman of the Year in 1975.

CLARENCE: It's hard to imagine that when Marie was born, she was so little they put her in a shoe box. They put some cotton in there and set her in the warming closet above the woodstove. It was in February, at a sawmill camp up Lightner Creek, just outside Durango. And in 1975, she was Colorado Woman of the Year.

MARIE: It was mostly for the work I did at the State Institution for the

Mentally Retarded in Grand Junction. I was a psychiatric technician and a dorm supervisor. I loved working there.

I also got the award because I guess I also saved these two peoples' lives that winter. I saw an old Mexican couple having car troubles on the road leaving Ouray to go to Silverton. When I started to help them, I noticed their gas gauge was really low.

They said they were planning to get gas in Silverton. I said I didn't think anything would still be open there. I recommended they go back to Ouray and spend the night, but they didn't have any money to pay for a hotel; so I gave them some money. And I gave the woman my boots, because her little tennis shoes were soaked through.

It turns out the pass to Silverton was closed that night. They never would have made it. So I guess I helped them out. Their children wrote into the paper about it. They said their parents would have frozen to death if it weren't for the "Angel of the Mountains." So that's how people heard the story.

Clarence running a mucking machine at the Hardrockers Holidays, 1975.

Clarence and George Davidovich (Marie's son), drilling at the Hardrockers Holidays.

CLARENCE: I was in Grand Junction on that Sunday in 1978, when Lake Emma caved in the Sunnyside Mine, where I had been working. I came back to go to work that evening, and Marie's oldest boy came running down and told me that the mine had caved in. The stream through town was running full of water, mud, and everything. It drained out of the tunnel. Tore out everything.

What caused the flood was that they mined up under, and they got too close to the bottom of the lake, and it just gave out. There was a lot of gold in it; and, of course, they wanted the gold, but look what it cost them to open it up.

There was a guy outside that was a caretaker out there in the buildings. He said when the lake water blew out the tin on the portal side, he thought it was a jet flying low. I guess it went almost across the canyon. It had a lot of force. It's just lucky a lot of people didn't get killed, but nobody was working on Sunday.

On the main level, we had a big hoist type of thing set up, and the cable went all the way up. The flood knocked all the timber out of that, and it blocked off everything. If you had been in there and survived the flood, you might have had to stay in there and just die. There was no way out.

But they spent a lot of money to re-open it again. They went all the way in, cleaned it all up, put new timber in. It's amazing they cleaned it up.

I stopped mining in 1991, when the Sunnyside finally closed. I was kind of a shift boss, or something like that. They had an underground warehouse. There were a couple women working in there, and I helped the women clean it out. Along with the hoist man and another guy, we were about the last people in there before it closed down.

I liked mining pretty well. I got to where I could do pretty well in there. When you had a decent contract, the money was good, but I don't miss mining anymore. I'm too old. I worked about forty-six years in the mines. It was a lot of hard work.

I never got very sick from working in the mines. I did something that a lot of guys didn't do. I made sure that I kept the dust down. I always had a water hose and blew the dust down. I don't have "miner's con," which is silicosis. But everybody I worked around, there's not very many of them alive anymore.

Walt Orvis

Ridgway, Colorado

My older brother acquired a mine around 1942, when I was about twelve, so I became a twelve-year-old miner. It was cold, dark, dirty, wet, miserable, and unfriendly. I didn't enjoy it at all.

Years later, when I went to work for Idarado, I wound up having the same sensation as I had when I was twelve. Every time I went into a strange place in the mine, I felt like my guardian angel wouldn't be able to find me. I had to put in at least two shifts in a specific spot before I was comfortable. After two shifts—looking around, looking up, checking it out—I was fine.

I started out with a partner named Frank. We were green, but we did pretty well. Even though it was just flat cold back there, we never shut down to take a break. We always kept a machine running. One guy would run down to the dog house to warm up. He would get a cup of coffee, get over to the electric heaters, and sit there and just shiver and shake. After a little bit, as he warmed up, he'd gulp down another cup of coffee and a candy bar and back up he'd go. Then the other guy could take off and run to the heaters.

The swing shift felt warmer. The swing shift was from about four in the afternoon until midnight. During swing shift, you had better compressed air, because there weren't as many people underground. There darn sure weren't as many bosses around, so I preferred it. I didn't mind the first week of graveyard shift too bad. Graveyard shift was midnight until eight in the morning. But by the second week, you were tired. Tired of all the work. You were just tired all the time. Graveyard was not common in the mines I worked in.

It takes a few workers to actually create the attitude of a mining outfit. If you have a happy outfit, it's only partly to do with money. It's mostly due to whether or not you like the other guys.

Sometimes you get a little animosity going, a little battling. There was always competition. It was an incentive plan, a way to make more money. One of the partners I worked with was Chuck. He was a hard working man—a working fool. Boy, he'd try to outwork the others. I used to give myself nine minutes to eat lunch. He and I would visit for coffee, cool down, and then he was up and going again.

I had to have a real pressing reason to go into another man's working place when they weren't there. When you're working the swing shift, and you see somebody strange coming into your area, they were probably in there to steal equipment. If I didn't have a specific reason to go into someone else's area, I wouldn't go. It's like walking down a street in Manhattan. You don't saunter down the damn street just rubbernecking. You go some place for a reason.

Gold was not worth a lot then, but if I did take any from in there, I gave it all away. All except one chunk, which I gave to my wife. I had a partner, Albert, who took some high-grade and brought it home. About two weeks later, he brought it back. The guys he was riding with said, "You're taking that in the wrong direction. What are you doing?"

Albert said, "Well I'm taking it back. I don't want it on my conscience. But I'll show you where I put it."

I worked with my son, Ken, for ten years, the longest I ever worked with anybody. We were a pretty damn good pair, and we worked opposite two of the top hands. We'd come out staggering, as they went in. I always said I couldn't even get along with Jesus Christ if he was on the opposite shift, but we got along with these two guys. They were good hands. One of them was Rick and the other was Dee. They knew exactly what they were doing.

One day before Christmas, we were the last day shift, and we got this red-headed guy. My partner had seen him before. That red-head could pick up one of those rails by himself and set it up. That's the kind of a guy we wanted for a partner. We finally got him that day before Christmas. I was mucking. The mucker ran on rails, back and forth, and it had keepers to prevent it from jumping off the rail. The red-headed guy was just signaling and helping when one of the keepers caught him on his toe.

I didn't know anything about it until we were outside, and we were in the bar. He stomped his foot and said, "I think it's broke." He told me he was afraid if he had hollered, I might have pulled the wrong way, so he kept his mouth shut and let me grind that metatarsal bone on the side of his foot, and he never said a word. Not a word going out. Not a word until we got in the bar, and he got a few drinks in him, and he was stomping on the ground saying, "Yep, yep, it's broke all right." He was a tough red-headed son-of-a-gun. He was back to work the day after Christmas.

I've had several close calls. One time I tangled up with a motor, crushing my pelvis. Another time I was on the skip, the elevator. It was going down a hole, and it came loose. A young guy was on there with me. It dropped us and then—bam! He bit his tongue. I drove my heels up through my ankle bones. They call that a "ladder heel."

My partner, Pete probably saved my life one time. There was a slab coming off. He gave me a shove. I thought, "What the hell?" Then— wham! He had seen it move out of the corner of his eye and gave me that shove. Otherwise I would have been right underneath that son-of-a-gun.

A lot of guys kept me from getting messed up pretty bad. I have been blessed with working with some of the best hands in the San Juans; some of the best partners. My son was, at the very least, one of the best.

I still dream of being in a mine. Nothing scary, just crazy things. So I guess I did like mining.

Rod MacLennan

Colona, Colorado

I was born in Ouray in 1931. Saturdays and Sundays, you would go down the street with your parents, your grandma, or someone. The bars were wide open. Miners were drinking and staggering up and down the streets. I thought of miners as a pretty hard, tough brand of people. I don't know whether I admired that or not.

My dad was a carpenter. I worked with him while I was in high school. When I had time off, I'd go up to the high country, hiking and just generally messing around. I loved it up there.

Then I kept hearing stories about people making four dollars an hour up at the mine. At the time that was big money. So at fourteen, I went up to the Camp Bird Mine and rustled a job. I got paid twenty dollars for my first week there. For a young guy, twenty dollars went quite a ways in 1945. But I didn't last more than two weeks because they found out how old I was.

So I went to work up at the Idarado, on the outside. They only had that age restriction for working underground. I worked on the tailings pond, which is the waste from the mill. You had to hand muck the tailings to build a dam. The waste runs through those pipes, which have holes in them, and you get these ponds. It's like a real thick milkshake. When it dries, you muck it up and build up your dam.

We were working on this dam, one day, when we heard a funny noise and things started moving. Then the dam broke. We just happened to be close to the wall and fast of foot. But there was a guy—I think John was his name—and the pond waste hit him. It took him down into the canyon. He was going down, hanging onto a pipe, and he turned around to look at us as if to say, "Help me." It was awful. They finally found his body about three miles down the stream.

When I first really went to mining underground, it was at the Idarado. They had a real nice boardinghouse there. They had a commissary, pool table, and all that. You could buy your clothing there, and this that and the next thing; and they really fed some nice meals. The meals were everything you wanted, and plenty of it too.

I was still pretty green, so they started me out as a nipper. I nipped for a raise-climbing crew there. I didn't think about being nervous, because you're amongst other guys. But I've seen people get claustrophobia. They hire on and ride the mantrip into the mine. They get out at the station; and, next thing, they were looking up the shift boss. They wanted out of there, right away.

I worked with one guy, John. He was an old miner. They called him "Old Chop," because he talked, talked, and talked all the time. He was scared to death of ghosts in the mine. If he had to be by himself in there, he just went frantic.

I remember, one time, I was pushing the tram. We used to foot tram in there. I was younger and pretty husky, so after we loaded the muck car, I'd push it out. When you pushed it back in, if you stood up, the timber would hit you. So you bent over going in there. That fan line at the end was going whomp, whomp, whomp. I got part way in there, and I saw this light coming staggering and running. Here comes Old Chop out of the mine. He was about six-foot-three. He was scared to death because he'd heard that noise back there and figured it was something going to get him. He was really spooked when he was by himself in there.

You always had guys that were jumpy and goosey. I remember this one kid, Bobby. He was goosier than the devil. Sometimes guys would save a piece of wax paper from lunch, and then on the mantrip going out at night, they'd light that paper and flip it over on Bobby. Oh man he'd go crazy, jumping up and down, and batting at it. That nervousness was just in his system.

But there was a lot of stuff to get nervous about. We saw a lot of accidents, and none of those mining accidents are very pretty. Usually if you get hit by a slab, you get smashed up pretty bad, and there have been guys killed motoring through air doors. The muck train would come, and the door was supposed to have opened, and it didn't. Of course that wiped the driver out. They'd go right through it.

All that machinery and that rock. There isn't anything soft in there. If you're not careful, there's bound to be accidents. One accident, I knew the kid. He was just a little guy, probably about five-foot-three. It was up at the Revenue Mine. I wasn't there at the time, but it was sickening to visualize. He was bringing a muck train out. Some way or the other, he fell out of the cockpit on the motor, and fell down into the track. The way they tell it, the train caught him when he was bent over, and that old train went right smack over him and just folded him together. He was mashed. That had to have been pretty gruesome.

I had a run-in up at the Camp Bird one time. We were timbering down a raise. The other guys had drove the raise, and they had driven it bigger than what it should have been. And the rail that hangs off the hanging wall, that your climber machine runs on, they'd used rock bolts instead of the regular climber bolts. Over a period of time, of going up and coming down, those bolts got to wearing.

Dave was my partner then. We started at about 500 feet up, off the bottom of the level down there, and we got down to about 350 feet. You timber down to a certain spot, and then you have to go back and take more rail off. You took rail off as you went down.

Dave started sawing one of those bolts off, and I could see it really had a lot of tension on it. I was yelling at him, "Dave don't cut it!" Well he'd just cut far enough that the head snapped off, and six pieces of that rail broke loose from the wall. The slabs came off the hanging wall, and threw us down against the foot wall. The climber machine broke loose from the wall and fell. The only thing that kept us from going down was a half-inch piece of metal on the back of the climber rail that didn't break.

So there we were, 350 feet in the air. We had a nipper guy down below, but he couldn't hear us because that slab had cut our air hose, and it was blowing, making a heck of a sound. We had crayons that you use for marking your timber. We wrote notes on wooden wedges and dropped them down, hoping that one would bounce out of the mucking compartment, to where the nipper was.

Finally the nipper guy started wondering what kept coming down. He had the nerve to take a chance and crawl up in there. Our notes said, "Shut the air off." When he did that, we could talk to him. We had to talk real slow, because your voices bellowed in that hole, but we were able to tell him that we had trouble up there.

We climbed back up to the top, by the station, and let the shakes get out of us. My legs were quaking so bad I couldn't hardly stand up. We realized we couldn't get out going up. It was an old workings. There was no way out. The only way we had to go was down.

Back down we went, and here's this climber. It weighed close to two tons, and it laid clear over against the wall. Luckily we had come-alongs with us. They're like a hand winch, with a cable and a ratchet.

I told Dave, "That's our only chance. We've got to try to pull that rail back up in place." There was a pinhole right about where we needed it, and we hooked the come-along to it and then ran a chain down around a piece of the raise climber. Here I was standing out over the open space there, 350 feet up, hanging on with one hand. I was just able to use my left hand to winch with. I started winching; and, I don't know, it had to have been the adrenaline building in my system to give me that extra strength, but finally that climber started moving. It came real slow. You didn't know if it was going to break or what. Then as it got straighter, it got easier.

We got it slapped up against the wall. Well then we could see the break, down there in the rail. Dave crawled down in it. He had a safety rope. He got down there and took some shooting wire, that you blast with, and tied it to the controls. Dave got back up into the timber and pulled on the wire. Well that released a lever, and the climber started going down. We had to jump down probably seven feet off the timber onto the top of that deck. We jumped and crawled inside the cage, and it coasted down to the bottom.

This was over a period of about four hours, and nobody knew what was going on. That situation there was the most scared I'd ever been in mining. I wouldn't go back up that raise. Something like that happens, you're not too anxious to go up and let it happen again. I figured it was a sign.

I went down, and I told the Camp Bird boss, I said, "I just don't want any more of that. I'm quitting." The next day I went over to work at the Standard Metals Mine in Silverton.

Standard Metals was a heck of a gold mine. Some of those people, they risked their jobs and everything else to go in there and find a few pounds of that ore. Most of the times they would come out with way more than a few pounds. Some of those stopes were pretty accessible. I've seen those guys at Standard come out of there with powder boxes filled with gold.

Rod MacLennan at the Camp Bird Mine, circa 1980.

My mining career mainly consisted of driving the drifts, and in a drift, you don't see that much ore. I got into a few places that had some. It was really pretty. Beautiful stuff, but I wasn't as set on getting it as some of the guys. If it was there, I'd take it. If it wasn't, it didn't bother me.

There was one night, I was driving drift, and I came out to the station. Gene and his partner—it may have been Rick—they came out to the station, and they had six or seven powder boxes chock full of ore. I said, "How the hell you gonna get that out of here, Gene?"

He said, "Oh, we made a deal with the muck train driver. He's going to take it out there and put it in the back of a pickup truck." I heard, a few days later, that they had sold those six or seven boxes for $3,500. Pretty good money for a night's work. That was the main thing for us—making that extra money.

An incident happened in the late 1970s at the upper level of the Henrietta Mine on the back side of Red Mountain. Ron Williams was my partner. Jerry Godbey and David Garcia were on the opposite shift. David and Jerry were day shift at the lower level. Ron and I were working night shift at the upper level. Both crews were driving drift from the surface. We were using a nine-to-twelve loader to bring in timber from outside. The loader was new and very dependable. Both crews had driven a lot of drift with it, and there had never been a problem.

The night of the incident, we picked up the loader at the lower level, where Dave and Jerry had used it on day shift, and mucked out our round from the night before. We got ready to timber. We dug a hole down to solid rock on each side of the drift. We put in our half of the square set, dug our holes for the other posts, measured for timber, and went back to get the loader to get the rest of our timber set from outside. The loader was back about sixty feet from the face where we were working. The loader would not start. We tried everything we knew to get it started. We tried this for fifteen or twenty minutes. No luck.

In contract mining, you are paid so much per foot or so much per ton; and if you don't finish your round, as in our case, you might as well have stayed home. Leaving a note on the loader for Dave and Jerry and trying to start the loader again, to no avail, we left for home around 12:30 in the morning.

At 7:00 a.m. the next morning, Jerry and Dave went looking for the loader, since we usually drove it back down to their level. They went into our drift; and there, in front of the loader, was thirty plus feet of rock slabs, rock bolts and mats, all on the floor of the drift, right up to our start of timbering. Maybe 80-to-100 ton of rock had come out of the back above where we would have been working to finish our timbering. And the thing that makes Ron and I feel, to this day, that God caused the loader not to start is, after Dave and Jerry looked over our mess (the cave-in) that next morning, one of them jumped on the loader, hit the heat warm-up button and turned the switch. The loader started right up and never failed again.

After thirty-five years, mining started to wear a little thin on me. It wasn't as much fun. It became harder. I was sixty-two when I stopped. I'd had enough of it by then. When you're young you think you're going to be healthy forever. You'll go on in there, and go to work, and let the devil take the hind most, but mining was just wear and tear on your whole body.

My hearing is gone. The air drills, those jack legs, and those stopers. They're nothing but noisy. My lungs are in pretty bad shape too. You eat that dust. I have second-stage silicosis. I can't go up into the high country very well anymore, because I can't breathe up there. I don't like that because that was my favorite country, up there.

But it's all those guys that got killed; that's what really hurts. I remember when a friend of mine came up and told me my friend Johnny had been killed. Johnny and I used to go fishing and hunting together. He was younger than me by quite a bit. It bothers you.

Joe Mattivi got killed the other day at the Grizzly Bear. I knew Joe since he was a baby. He was also born and raised in Ouray.

All of them have their effect on you. You feel it.

Mable Lyke

Grand Junction, Colorado

I was born around 1932 in Nebraska, in a tent somewhere near Lincoln. The courthouse burnt down, so I have no birth certificate. All those records were wiped out. All I remember is that my parents lived in a tent, and they worked a truck garden with vegetable stalls.

I got married when I was nineteen years old. He was a long-time coal miner. We moved up to Silverton in 1962, so he could mine the hardrock. He worked mostly at Standard Metals. They wouldn't hire women, so I drove a school bus. I worked in cafes. I did everything I could. We had two sons, and at that time they were just little bitty creatures.

In 1970 my husband went to work in a coal mine over in Redstone. I bought an old school bus that had been converted to live in. My sons and I moved into that and drove it from Silverton to a little town near Redstone. Eventually we moved into a house there.

I made $100 a month driving a school bus again. There were about seven kids in our little town. I would go pick them up and take them down to Redstone to meet the big bus. Then I met the big bus at night, and I'd take all the kids back home. All I had was an old little van-type thing. It was worthless. As soon as it hit snow, I had to chain the tires. That was practically every trip I made.

∞

I started to buy and sell coal on the side. At first I just got some for us. But it turned out, people all around us needed coal, so I started to sell it. I had a three-quarter ton pickup then. The boys and I would go over to the Bear Coal Company in Somerset. We'd load the lump coal off the ground into the truck, and then haul it back to where we lived.

Our house burned down in 1976. We lost our dog, cats, absolutely everything. I knew these people that had the motel there in Marble. They put us up in two rooms.

Long about then, the government said the mines had to have so many lady miners per male miners. The boss at the Bear already knew me from loading up coal, so he hired me. I became the first woman coal miner at the Bear. The boss took me over to begin my shift. He told those guys, "She's never been underground before, but you've probably seen her out here loading coal."

They all said, "Yeah we have," and then this one guy said, "But sometimes I use awful bad language in here." His name was Bob.

I said, "It's okay, Bob. I can't understand French anyway." So they thought that was funny, and I was more or less in. When my first shift was up, they asked if I was ready to go out. I said, "No." I wanted to stay. With mining, you either love it or you hate it. I loved it! I just couldn't get enough of it.

They put me on the belt line. I worked day shifts and swing shifts. Swing shift is when you go to work at 3:00 p.m. and you get off at 11:00 p.m. They'd all find one or another reason to come around and check to see if I was doing my job. I was working as hard as anybody else, so I became one of the guys. No problem whatsoever. They knew I was in there to work. When you're a woman, you're either a "worker" or a "dolly." And I was no dolly.

When my husband got itchy pants again, he moved on to Hotchkiss to mine there. I stayed on at the Bear. I would drive back and forth from Somerset to Hotchkiss sometimes.

My youngest boy decided he wanted to be a motorcycle mechanic, so he went to Florida to do that. My older boy, Buddy, had been in Grand

Junction a while, and he decided to come live with me. I would come off a day shift, and my dinner was sitting on the table. He had done the housework and the laundry. We had a fabulous time. We can be in the same room and visit with each other and not say a word. It's the same way with both boys.

I was the only woman in the Bear for most of the entire time I was there. They did bring in another lady, one time, and put me to training her. I remember she told me, "I'm not looking for a friend. I have friends. I'm in it for the big bucks."

I said, "Them bucks are gonna come awful hard," but I trained her the best I could. Well, soon she was crying on my shirttail. Oh, it was too dark, and she was scared of the dark. Oh, it was too hard labor, and she was tired. I didn't have much sympathy for her, but I tried to treat her just like people.

Shortly after she came, I wound up with female troubles. I started to hemorrhage. It wouldn't quit, so I finally went to the hospital. I wound up having major surgery. When I went back to work, the other lady had already quit.

I bid for a buggy operator job in there and got it. Then they had a breakdown up at the face, where they actually mine the coal. When they had this breakdown, they stopped mining; so they put me back on the belt line. That made me mad because, damn it, I had gotten off that belt line. I thought, "Let me help you up at the face!"

I let my temper get the best of me. I was shoveling stuff onto the belt—rocks and coal. This huge piece of rock fell off the belt. I grabbed it and tore my self loose inside. I wasn't strong enough yet from the surgery. I stuck with that damn rock, because I was so mad, and finally got it back on the belt; but that was the end of me, right there. I had to take a leave of absence. I never did get back on for the Bear.

Eventually I got on for Northern Coal, and I worked there until they shut down. There were no more coal orders. I tried to get back on somewhere else. I went clear up into Wyoming, but they were full up and didn't need anybody.

I would have done anything to go back underground and mine coal. It was probably the hardest job I ever had. It was black, and it was dirty; but I liked it down there. I was used to manual labor; and, here, the money was super good, and I had insurance. I thought that was a good deal.

In 1982 I was in a car accident. I got sideswiped, and it wrecked my neck. I took up carving rock, making little animals and such. My boys talked me into selling them at the various Mountain Man Rendezvous in Silverton and wherever. I told the boys, "I'll go, but I ain't wearing no dress. I don't wear a dress for nobody. If I can wear the leather pants and leather shirts like the men, I'll do it." So I made a leather pants outfit and started selling my carved rocks at the Rendezvous.

When I finally settled with the insurance company about the car accident, I bought my little house in Grand Junction. It rocks some and the windows breathe, but I had a house.

My husband and I separated. He hadn't quite moved out when I had a stroke. The stroke kind of sliced me in half, took my whole left side away. Now I'll be talking, and I'll just draw a blank. My left hand is really giving me static. I can't hold the stone to carve my animals. So my ex-husband stayed, as my roommate. He does lots of things around here, that I can't quite make work. Mop the floors and stuff. He earns his keep.

My oldest boy and his wife gave me their old computer. I run the keyboard with my right thumb, which works really great. I've always been independent. Took care of me and a dozen others. Now I can hardly take care of me. There are days where if I went out to the shed, I couldn't find my way back home.

But I just hang on. I'm too bull-headed and determined to let it go at that.

Dave

Montrose, Colorado
"Dave" is a fictitious name. He wished to remain anonymous.

We were raised in Bisbee, Arizona. There were open mines all around—old mining claims and mines that had been closed down. The people had just walked off and left them. And so, as kids, we'd just go in exploring. We weren't supposed to, but I don't think we broke any laws going into them. Luckily we never got hurt.

I got interested in mineral collecting. There was a barber who had a fantastic collection of rock specimens. The miners would trade specimens for their kids' haircuts.

The first time I went to work in a mine in Colorado was 1957 at the Idarado. It was a big mine that went clear through the mountain. Part of the men worked from the Telluride side, and part of them worked from the Red Mountain side.

I started out as a mucker, getting the piles of ore and stuff into the muck tram and out of the mine. Then I started mining. Mining was kind of an advancement. One thing about mining is that the bosses tell you what they want, and kind of leave it up to you to get it done your own way, within reason. You don't break safety rules to do it, but you may cut corners. As far as drilling a round out, you do it your way, as long as it breaks. It's one of the things that makes it nice to be a miner. You work for yourself.

If your partner didn't hold up his own end of the work, you made sure either he quit, you quit, or somehow you got split up; because work equaled money. I had a lot of different partners, all of them good partners. Most have been dead now a long time. I lost one recently to lung cancer.

One partner I think about is Curly. He was quite a drinker. He'd always go out and celebrate one thing or another. He didn't show up one Monday. Next day the boss said, "Curly, where was you at Monday?"

"Boy, God," he said. "It was my birthday. I celebrated."

The boss said, "Well you better go up to the office. The old man wants to give you a birthday present." They gave him his last paycheck.

The thing a lot of people don't understand is, it was a good time for mining. There was a lot of work. You could go to work any day you wanted. I mean you just decided you didn't like that job no more, and you packed up and went some place else.

I lived here in Montrose, no matter what mine I worked at, and commuted. I left my wife here, and I'd go wandering around. But when I went to work at the American Tunnel in about 1959, we all moved over there to Silverton.

We were going to stay in Silverton, but some of the guys and I got caught up in trying to promote a union, and we got run off. That was about 1961. We were "a reduction in force." It was the polite way of getting rid of us. There was no trouble. The miners just voted the union down. It was mostly guys from Silverton who didn't want it. That was a funny place, Silverton.

One time, about 1970, a huge storm shut down the highway to Silverton. Couldn't get in or out for four or five days. At the same time, they also had a slide run between Silverton and Durango. They couldn't get any food or milk in for the kids, nothing. But that didn't seem to worry people until they got down to only thirteen cases of Budweiser in the whole town. That's when people really started worrying. Silverton was a big Budweiser town. So they put all their resources into clearing the Durango side and finally got that road open.

People in Silverton kind of didn't like outsiders. I never had any trouble in the mines, but our family life after, you didn't know anybody. It was hard to live up there. After the union deal got me "reduced," I went and got a job the next day at the Camp Bird, south of Ouray.

I got hit bad with a slab once. That fractured a vertebrae, but I went back to work within a month. Then one time I was drilling when a rock came down and hit me on the foot, broke that, but that's probably about all. Broke an arm, broke a hand.

But I think some of the rides coming off the hill were worse than the working conditions. A young guy got caught in a slide, and it took him off the cliff up there. We dug for him, but he was dead by the time we finally got to him.

If a slide ran, they wouldn't let you walk over them at night. So you had to stay up there at the mine all night, until it got daylight. You'd walk down as far as you could, until they could get a truck up there and haul you out. I spent several nights at the mine because of slides. We'd stay in the change room.

There was one time we were coming home from work, during the day. Bobby and Ike were in the back. My old Chevrolet had a banged-up door in the back that didn't work too good. I was driving. Joe was in the passenger seat. We were coming down that first big curve and, oh, it was slick. I was really going slow, but that car it went—shoo! It spun around once. Joe opened the car door. I yelled, "Jump Joe!" So he jumped. About that time, the car spun and hit him in the rear end and knocked him off the cliff. We kept sliding over to the edge of the road, hit a berm of ice, and stopped. I got out and here comes Joe climbing up the hill. He had one boot on. He said he lost the other boot. We all started laughing at him. He said, "Well, laugh you damn fools!"

That's the last day Joe worked. He quit and never did go back up there again. He wound up having to get hip surgery. He swore it was from that fall he took over the cliff, but I don't know. It was only a forty foot drop.

You had so many friends in the mines, and you'd all joke and then go back to your work. At the end of your shift, when you were cleaning up, one of the favorite things was to wait until everyone's face was soaped and then run around turning off the hot water. Listen to them squeal. That water was cold. There were a lot of practical jokes. Turning out lights. Little things that were really harmless.

We had a few bar fights. I used to fancy myself a fighter. But they had to make me pretty mad first. There was a lot of arguing about who was the best, who was the worst. With contract mining, the more money you made, supposedly the more work you had done, and the better you were. But pretty much everybody worked hard, and everybody made more money than you could make in town.

When I started collecting rocks, that made it a whole lot easier. I collected specimens and sold a lot of them. But it wasn't only the money I was after. They were pretty. It wasn't really high-grading because quartz and the different minerals don't have much value in them. The gold does. That's the real high-grading; the gold. They didn't want you packing off any of that, but a good rock specimen, I don't think they cared much.

On our days off in the summer, we did a lot of fishing, up at Blue Mesa. Take a case of beer, go up there and fish. Boy, it was good fishing then. But in the winter, there was a lot of spending time at home.

One time some man came up to Silverton and said he was looking for a Gallegos kid. This lady, Mary, told him, "My God, there's forty some Gallegos children in school here. How do you expect us to know which one you're looking for?"

The guy looked at her, and he said, "What do you people do up here in the wintertime anyway?"

She said, "Well, we don't fish!"

So I guess you could say that my life mining has been very uninteresting, but I had a good time doing it. And if it was here again, I think I'd do it again. If I were a cash register person or something like that, I'd just go crazy on it. I don't know how those people can stand it.

Brian "Buckwheat" Elliott

Montrose, Colorado

*M*y Dad moved us to Telluride when he went to work at the Tomboy Mine in 1941, during the war. I was eight years old. We lived up there on the side of the mountain. They'd do some blasting in the dead of night, and sometimes a rock would hit the roof. Us kids would wake up and cry.

I had a black freckle on the end of my nose. That television series came out—Our Gang—and there was that one black kid they called Buckwheat. So people started calling me Buckwheat, because of my dark freckle. And that's the way it's been all my life.

When I started school, my folks got separated. Mining wasn't my game, but I had to work. I started out in the mill. I went underground when I was about eighteen, up there at the Idarado in Telluride. I was young and had a lot of disagreements with the company. I was always wanting more money, and they were always wanting more work. I'd get suspended and brought back, then suspended again.

Right after the war, there was no problem getting a job. I could get fired from the company in the daytime and go to work somewhere else at night shift. The main thing that they wanted then was work. I believed in working hard, but like I said, I was kind of ornery. I'd get thrown in jail sometimes, after a little dispute downtown. An argument over who could drink the most, I guess. Who could out mine the other guy. I'd a little overdo it. I was always on the fight.

∞

There was an old fellow people called "Whispering Jim," because he talked so loud. He had a lease from Idarado on some of the old Telluride Mines' ground, which they had mined years back. I worked for him about three months.

He got caved in one time. Boy, I never worked so hard in my life, trying to get air to that guy. He had two other guys on a different level, driving drift. I went and got them, and there was another guy close to where I was. The four of us dug as far as our arms could reach alongside Jim's body and got the air there.

We could hear him, just weird sounds. It took us about four hours of digging, but we finally got him out. His eyes were bloodshot from the pressure, the weight and stuff that came down on him, but he was okay. He was a tough old devil.

I went back to Idarado and talked to the foreman. I said, "I'm not gonna spend a week or two coming out here every day, asking you for a job, and not get one." I said, "You're either gonna give me a job or you're not." So I went back to work for Idarado.

The company's main goal was production. I don't know if they let it jeopardize people, but they did cut corners if they could. They liked young guys who were willing to take on bad areas.

You had to be careful, because nothing gives in the mine. It's all solid. Your equipment is solid. Your explosives are sharp. So you got to think a little bit, you got to have a little common sense. You have to learn how to take care of yourself. It's a really unique profession in that way.

Most all the miners, they had come up through hard knocks; and everybody was there for the same purpose—to make a living. You had a partner or two partners, and the harder you worked, the more money you made.

It was hard to find six men who could work together. There were always the ones you just look at, and you didn't like one another. And there was always one guy who thought he was doing more work than the other guy.

Brian Elliott (far right) and other miners underground.

I was pretty lucky. There were five other guys and me, and we all got along. We did a lot of stoping up there—we took up thirty-seven stopes. You had to be pretty lively, because the rock talked to you a lot. Popped and cracked. We worked together about seven years. They were real good friends of mine: Cecil, Don, Les, Leonard Williams—a good hand, he was my partner—and Hank, but he died.

In one sense I liked mining. In another sense, I didn't. When I was stoping with them guys, I liked mining because we were getting along good, and we were making top money.

We had a walk-out strike, when us guys were together. We were making big money, and the supervisors got the idea that we were writing our own ticket, which absolutely we were, and we had a right to. I mean, they wanted the tonnage, and we were giving it to them.

Well, when they had the walkout, it was because the supervisor from the Red Mountain side gave me a three day suspension for not wearing my safety glasses. I had been sitting in the dog house—it's the little room in the mine where you eat your lunch. Anyway, he gave me a three-days-off for not wearing my glasses in there.

Of course the word got out right away, all over town, that they suspended me for not wearing my safety glasses in the dog house. They had been suddenly writing a lot of three-days-off for safety violations, trying to make the company look good for OSHA [Occupational Safety and Health Administration].

Brian Elliott with "Sally."

Anyway, I came out of the mine that night. My three partners who were getting ready to go in for night shift—Cecil, Les and old Hank— they started talking to some of the other guys. There were lots of problems anyway. My suspension just kind of gave them an excuse, so they walked out. The night shift didn't go to work that night.

Eventually we all went back to work. Got our jobs back, but they wouldn't let us guys work together again—separated us. Cecil, they put him in supervision. Hank, they put him in safety supervision. Me, Willy, Les and Don, they split us up to different parts of the mine. Don went down the road, and he quit. He went to work in uranium. Willy quit. Les quit.

They shuffled me around here and there. They got me pulling pillars. Pulling pillars is one of the most dangerous jobs. When you're stoping, every once in a while you leave a twenty-five-foot chunk in between, from wall to wall. You went over and under it. It held the walls. Some supervisors would get a little carried away with them pillars. It cut your tonnage down, so you'd pull some of them.

I finally got to drive drift, which is what I always wanted to do. There's something about driving drift. You think maybe one more round of explosives and we'll hit it—that big piece of gold. It kind of gave miners the incentive to keep breaking ground.

I was driving drift when a bull hose came loose. It's an air hose hooked onto a mucking machine. It's got 110 pounds of air pressure. When it came loose, it was just like a whip. All of this air coming out of it, blowing rocks and dirt, and everything. It hit me in the face and knocked all my teeth out, what teeth I had left, and broke my cheek bone.

I was fortunate enough that, subconsciously, I knew if I got to it and held on, it would throw me ahead, out of danger. By crawling toward it, I could get away from it. I did that and wrapped my arms around it. I could feel the pressure. I put all my weight on my one foot and turned loose, and it just shoved me up on the muck pile—which was what I wanted.

My partner had took off running down the drift, because he'd heard this loud bang and knew what had happened. Anybody who works with those air hoses would know. Anyway, he ran down the drift about 300 feet and shut off the air valve. When he came back, I was sitting there on the edge of the drift, pulling teeth out of my mouth.

He put me on the motor and took me to the hoist room, probably about a mile down. We got there and called up the shifter and told him I was hurt. The shifter said, "Well, how bad?"

"He's pretty bad," my partner said.

It was about 2:30 in the morning. Blasting time would be 3:00 a.m., and the man trip would usually go out then. The shifter said, "Well can he wait until the man trip?" So we waited until the man trip went out, and I was still holding my mouth, with blood running all over everything.

The day-shift foreman took me home. My mouth was still bleeding, but I didn't want to wake up my wife. So I sat there at the kitchen table with a quart fruit jar, and I kept spitting in it with the blood coming out of my mouth.

My wife must have known something was wrong. She got up. The jar was almost full. She said, "Well what do you want to do? Do you want to see the doctor here, or go right to the hospital?"

I said, "Hell, I don't know. I'll just sit here, I guess." I finally agreed to at least see our old family dentist in Ridgway. This was like 4:30 in the morning. He said to bring me on down. The dentist shoved my mouth full of tea bags, and he sent me to the hospital. I was there a couple of days.

I never talked too much about it. It seemed just one of those things that happen. But later on, I got to hearing some of these rumors that were pretty upsetting—like that the supervisor damn well should have paid more attention to what had happened to me and should have gotten me right out of the mine. The more I thought about it, why, it just made me mad; but I went back to work and pretty soon that was all forgot. It was just another thing.

The main problem they had there was too much powder smoke. The shifts coming in together, right behind one another, didn't give the mine time enough to air out. Finally they got their ventilation system going, but there were some awful bad places in there that were real smokey. I think it affected my health. I don't have hardly any lungs left.

When they brought diesel equipment in the late 1960s, that added to the gas and powder smoke in the mine. It definitely had its effect on the whole damn mine. There were some people just a little bit more susceptible to it than others.

Most guys only knew one thing, which was right there at the Idarado. That was all right if the Idarado never shut down, but it did. And that put a pretty tough deal on everybody, because most of them, like me, thought you'd have a job there all your life.

I went up to that open pit mine, over there at the top of the mountain. I think it was called Summitville. They used a different process. They leached. They had huge equipment up there. You moved a lot of rock, and then it leached down through these mats and a watery substance. From there they pumped it up into a closed-in, high-voltage electricity area which pulled the gold out of the water.

As I understand it, Iranian people owned the ground, so to get around some of the environmentalists here in the states, they let a Canadian company do the mining. The Canadian company, to get around some of our mining rules, hired a construction outfit out of Montana to do the work. I worked for them.

We worked up to sixteen hours a day. That's when my health started failing me. It's 12,000 feet up there. The weather wasn't all that great, and they leached that ore with cyanide. A cyanide spray.

Your equipment, machinery, was supposed to be cabbed-in when you were out there. Of course they'd think you could get by without a cabbed-in truck or Cat. I didn't have any complaints, really, but then I got a nosebleed one night.

I couldn't get my damn nose to stop bleeding. I called up the boss and told him that I was going home. I went to Telluride, where my wife was. Of course she was kind of scared. She brought me down to the Montrose Hospital. They packed my nose, and they finally got the bleeding slowed down. They thought they had it, and then all that packing they'd shoved up in there, popped out of my nose. All the pressure in there, just brought it out. So they started working on it again and finally got it stopped. I went on home, but I didn't feel too good for a little while after that.

From then on I've had problems. I never was released to go back to work up at Summitville. I continued to have breathing problems; and, I don't know, just problems.

I went to the Veterans' Hospital in Grand Junction. They were real good to me, and I had one of the best doctors, I think. She was a woman, and she did more good for me than any of the rest of them. I had a tracheotomy. With that and medication for depression, she got me going pretty good. I still go down there about once a month.

As far as my lungs go, I can't do anything about that. I mined twenty-five years, but I didn't work for any company long enough to be able to file a health claim. I know damn well my lungs caught their death at Idarado, but they said that because I was still able to work when I left—because I went on to work at Summitville—Idarado wasn't responsible.

The truth is, I was in no condition to work when I left Idarado. I just worked because I had to. I was fifty years old, no where near retirement age, and you just go on and do what you have to do.

Benny Salazar

Interviewed in Durango, Colorado

I was seven years old when my dad went into milling in Silverton. He loved the mill, and he got killed in the mill. The final ore bin was where they crushed the rock real fine. It's a big bin, full of ore on the sides of it. It would freeze in the winter. They'd have to go in there and pick at it. My dad didn't have his safety belt on when he was picking on it, and it all gave in. It sucked him right through there. They got him off coming out of that bin. I never saw his body. I was too young.

When I was in school, I met this man who owned the Brooklyn Mine. I did hand tramming for him in the summers, me and his kids, outside the mine. I was twelve or thirteen years old. We were just kids having fun.

The first time I actually went into a mine, I was fifteen years old. It was with Bill, the superintendent of the Highland Mary Mine near Silverton. I'd go with him during the summers and spend quite a few days with him. He showed me how to mine. In those days you didn't have regular electric lights. You had carbides. It was dark and wet, but I loved it. I loved it ever since. I don't think I'd rather do anything else but mine. Every day was different.

One time I was driving a stope in the Idarado, and I hit wire gold. Wire gold is one of the most beautiful things you'll ever see. The other miners, I guess they could smell it. The timbermen, the trammers—they were all up there getting that stuff. They were fighting for it. You hear about gold, but when you see it, it shocks your hair. Gold drives a man crazy. For the money, I guess, but it's also the beauty of it. It's like butter.

If any miner tells you they didn't high-grade, they're bullshitting somebody. Even the bosses. The bosses probably got most of it. There ain't nobody in this world that goes underground, and mines hardrock, and doesn't take some of it. I'm not saying he's going to take it all, but just get some of it.

I was at the Black Bear, and I hit gold one night. A lot of it. That Black Bear gold was heavy gold. It had a lot of silver with it. I think I had about five or six powder boxes full. You can put it to the side, see, and then when the supply train rolls out, I put it on the supply train. He'd take it out for me. Then he'd put it in the bus, because we all ride the bus to and from work. When we got out, the bus driver said, "God damn you, Ben. You're going to get me in trouble."

I said, "Well it's going to be fun getting in trouble. What the hell!"

So off we go, boy. We ended up at his house. Started drinking, looking at the gold. Well, I forgot to go home that night. My old lady didn't know what happened. She called John, and there I was. We were both in trouble with her.

When you find gold, people who buy it know right away. The next day, hell, I had a line coming all the way from the Grand Imperial Hotel to my house. They'd buy it from me, fix it up and then they'd pop it over to Mexico. Sell it for more.

The gold I took, I gave some away, and I sold a lot. Good money in it, but you always keep a "pocket piece," they call it. Keep it in your pocket. But ever since I quit mining, I leave it in a drawer.

We used to raise hell in the mines. These guys, they used to like to go to the machine shop and bullshit after lunch. I was on the supply train one time, and I cut a broom handle off and made it look like dynamite. I stuck a fuse into that wrapper. I lit the fuse and threw that stick in the shop, where they were bullshitting. They started running, trying to make for the door. Oh they wanted to kill me when they caught me.

This guy we called "Mule," he played the same trick on me at the mill. That was a famous one. I was working in the crushing plant. I was on that shaker, watching what comes out. That evening I watched as, coming off that shaker, was a damn stick of dynamite. I was trying to grab it, so it wouldn't go into the crushers and blow up. I got it, and I was yelling to my boss. The fuse was getting shorter and shorter, so I threw it over on the floor at the bottom there. That thing had a firecracker at the end of it. Scared the hell out of me when it went off. That damn old Mule. He was always playing tricks on me.

When I worked at the rod mill, I had to take care of the reagents. Reagents are like acid. That's what they mix with your ore to clean it. Anyway, when the guys would go home, they'd have to come right under me. This one night they were going home, and I was working the night shift. I was up there waiting for them with a hose. I turned that hose on them, and they thought it was that reagents. That was a fun time. If you can't have fun on a job, why work in it?

I've been truly scared one time, at the Gold King. It's an old mine above Standard Metals. That's the only mine I've been afraid of. The water in there had arsenic in it. It eats up steel, nails, and everything. I had a mucking machine. I had lent it to them. They had it sitting in there for maybe a month. It was sitting in that water. When they came back to get it, it's got no wheels—the water ate all the wheels off it.

One time we were going up this big raise in the Gold King. I got off on the fourth level. I looked up and there was this two-inch cable hanging by a thread, 800 feet above. You could tell the water had been eating on it. I just froze. It went through my mind that if that damn thing came down, it would go right through me. That was the most scared I've ever been underground. I got hurt at other mines, but the Gold King was the only mine that ever scared me.

In 1968 I messed up my back at the Old Hundred Mine. I was riding a machine, and it drove me against the face. That machine knocked the hell out of me. Ruptured the discs in my back. Ever since then, I haven't been worth a damn.

I was off for four years. That's when I met Jim, and he asked me if I'd be a boss for him. I re-opened one of the oldest mines in the San Juans, the Little Giant. It was started in 1872. I opened it 100 years later, in 1972. We hit a little vein, six to eight inches wide.

I have a ghost story. People may not believe it; but, by God, it's the truth. I had a lease on a mine on Deer Park. I had my kid with me. We took groceries and everything. We had a cabin there. My kid had to go back to town, so he told me he'd be back in about three days. I said, "All right, I'll just be messing around here." I liked to stay up there. It's quiet.

That evening I went looking around to see if I could find old bottles in the portals of those little mines they have up there. Then I had supper, and I went to bed. Around about 2:00 o'clock in the morning, I started hearing this big ol' noise, like a chain hitting the floor about two or four times. I thought, "Well those God damn groundhogs." I thought maybe it was groundhogs messing around with those bottles I found.

But all of a sudden, whatever it was started going around the cabin, making this weird noise, like a moaning. Real strange. You can't describe the yell—this moaning yell. I had a gun next to me, and I kept thinking, "I ought to shoot that, but what am I shooting at?" I thought it might be somebody trying to scare me. It just kept going around the cabin. I

Benny Salazar (right) and his son, Andrew.

thought, "If something comes through that door, I'll shoot it." But nothing came in. The moaning lasted until about 3:15 in the morning.

It turns out that years ago, behind that cabin, they buried a miner's son, and the noises I heard had gone right by there and up that trail behind the cabin. The next morning I got up and looked around the cabin. No tracks. Another man, a friend of my sister's, had the same experience in the same place.

When my back got better, I went back to drilling and everything. I told social security to get me off of it. No one in social security thought I'd ever mine again, but I did pretty good for a while.

I went to work at the mill here in Silverton for Echo Bay. I was in the crushing plant. One time I had my back to this shaker, where all the ore comes through. The damn rocks were coming off of it, and a huge one threw off of there. Messed up my damn back again. Well, I've been down ever since. My doctor said he couldn't fix it. I was sixty years old. No spring chicken. Not worth an operation.

I started drinking pretty heavily. More than usual. The police probably did a good thing when they put me here in jail. That night they pulled me over, I had taken my son home, and then I was going home. I was drunk. When the state patrolman stopped me, I got mad and jumped him, and here I am!

Yeah, I loved mining. I never wanted to do anything else. The good old days are gone.

Pat Donnelly

Silverton, Colorado

I was flat, busted broke when I got to Silverton in 1976. I'd run away from home. I had just gotten a divorce. I couldn't get away from my ex-husband, and I figured maybe he wouldn't be able to find me here.

Running away scared me to death. Usually the farthest I ever got from my mother was twenty miles. I didn't know if I could make it by myself. When I found out I could, it changed my whole personality. I had been a very timid, shy person when I came here. It only took two years working in the bar to kill that.

Eventually my ex-husband went through my mother's address book, and he came up here looking for me, but it didn't do him any good. By then I was good friends with the law enforcement. They said, "If he gives you any trouble, just let us know and we'll take care of it." The law enforcement were real good about things like that. They looked out for you.

I worked at the Grand Imperial Hotel. I started with cleaning and waitressing, and graduated up to cocktail waitress and bartender. I became manager for a while, before I decided to work at the mill.

When I was growing up, you were supposed to go to school and, in the meantime, learn all the functions of being a wife or a homemaker. Learn how to do your crocheting and your embroidery and wash, iron, clean, scrub, and cook. I learned all that. And because of that, I'm actually considered, by a lot of people here, to be very old fashioned. It's kind of funny. People always call me to find out how to get rid of certain kinds of stains, and stuff like that. I didn't mind, but as far as making a living, working in the mill sure seemed like a better idea.

In those days you had to rustle your job. You didn't call up to the mill and write out an application. You went up there, and you asked for a job. For two months I walked that road, back and forth, from town to the mill, to ask for a job.

There was a transition at that time, a change in the mill superintendents. The one they had there refused to hire women. He always told me there was no work, to come back again. So I did. And I did. Then it changed hands again, and they had another person there, a temporary superintendent. He wouldn't hire women either. If you wanted to take it to court, you could, but when you're flat, busted broke, where are you going to get the money?

They finally changed over to another superintendent. I went up a couple of times, but I didn't get to see him. He was working. But one time when I went up there, I saw him. It was bright and early. He said, "Well, have a seat. I'll be back in a few minutes, and I'll talk to you." Then he took off inside the mill.

At 3:00 o'clock in the afternoon, he came back over and he said, "You still here?"

I said, "Yeah, you told me to wait, and you'd talk to me."

He said, "Well, I didn't expect you to wait all day."

I said, "I'm gonna tell you something right now. I've been walking this road for two months to get a job. Two of these idiots up here wouldn't hire women, and I made up my mind—you tell me you got a job for me, or you don't have a job. Decide one way or the other, because I'm not walking that road any more. My feet hurt."

He laughed, and he said, "Do you have a resumé?" I said I did, and I handed it to him. He went into his office, and he came out a few minutes later. He said he was impressed with my resumé, because I'd worked in factories most of my life. "But the only thing I got is temporary work. Maybe three days a week. No benefits. You'd be doing cleanup."

I said, "I don't care what it is. It beats making three bucks an hour." I started working in the mill in 1981.

∞

It was tough when I first went up there. Working in the bar, you tend to gain weight, and I was a little heavy. Also I never did work a three-day week, like he said. I worked seven days a week, ten hours a day, with a wheelbarrow and a shovel. I lost twenty-one pounds in twenty-one days. I thought I was going to die. It was funny.

My job was to take the wheelbarrow and go to an area where they'd had a spill during the night. I'd take the rocks and ore to the crushing plant, where it gets crushed into small rocks. From there it goes through different areas, getting ground even finer. The crushed ore then goes across a jig that vibrates, and your gold and heavier particles fall down and go to a different area for processing.

I was there almost four months when I was about ready to quit, because it was so hard. This friend of mine, who was the foreman, said, "Hang on just a little while longer." About a week later, the mill superintendent came walking through one morning and said, "Oh, by the way, have I told you yet that you're on permanent? Started a couple of weeks ago, but I just didn't get a chance to tell you."

Oh boy, I jumped up and down for joy. That meant full benefits, insurance, and everything like that. Plus it meant a small raise in pay and an opportunity to get off that wheelbarrow and shovel.

So I went up to being a grinding operator. That's where you got to watch the ore that comes through, to make sure you have the proper amount of water, and that nothing's plugged up, and that you have the right amount of reagents going in to clean the ore.

Reagents could be fairly toxic. I have small scars from it, but that was before we had safety clothing. Later on they made safety clothing available to you. If you got anything on you, and it went through to the skin, it was your fault. That stuff eats you alive. It's caustic. When you mix it with sweat or water, it starts burning like an acid. You can actually see it burn and bubble. It can eat you clear to the bone.

They use reagents to clean the metals that have been ground. The ore has to be cleaned to be able to adhere to the bubbles in the floatation devices. A floatation device is like a vat with a huge agitator. You add air to it, so the contents bubble and wind up looking like a milkshake.

The bubbles are special. They're sticky. The ground-up metals stick to the bubbles. The air forces them to raise up and overflow the edge. Then the bubbles pop and your lead, zinc or gold—that have stuck onto the bubbles—wash on down to be separated out.

This floatation idea was actually created hundreds of years ago when a Welsh washer woman was cleaning her husband's clothes in a washboard tub. She had the soap in there. She was washing the clothes; and, of course, the soap suds ran over the edge of the tub. Well, when the old man came home and had to dump out the tub, he noticed that there was a little ring around the bottom of the tub, on the outside, and there was concentrate there. From that little accident came this really complicated floatation circuit that we used at the mill.

The men didn't think women belonged in the mill. The work was too dirty, they said, and they didn't like having to watch their language. But eventually I became one of the guys.

I ended up talking like they did—cussing. I could sling them out just about as fast as any man could. Worst habit I ever had to break. Now I only tend to slip back into that habit when I'm drinking. I was telling somebody the other night, "I haven't lost my temper in over fifteen years. Don't you be the one to make me lose it tonight, because you're going to catch fifteen years of hell."

I also haven't hit anybody in years. Last time was my second ex-husband. We were married then. It was his fault. He hit me first—bad move. I believe everyone should have a little bit of a temper, because that means you'll stand up for yourself and your rights.

At the mill, we had to take a test on a regular basis for lead and mercury poisoning. They tested blood and urine samples. One time they found out my levels were real high, and they pulled me out for six weeks. That was pretty bad for a while. With lead poisoning, you get really tired, and you get aches and pains.

I wasn't supposed to go back to work until my levels got down within the right range, but I talked them into letting me go back sooner, because we had an idiot working in there, and he was losing a lot of the gold. He wasn't catching it. I told the doctor, "I have to get back in there. They're losing their butt."

He said, "Okay, but you only stay in there as long as absolutely necessary, then you get out." So that's what I did. They eventually got away from using mercury. Now they leach with cyanide. I'd rather play with mercury.

∞

Eventually I went to work in the mine. My boss told me if I could pack the jack-leg drill down into the drift, set it up, get the water and the air hooked up to it, he'd teach me how to drill. I worked very darn hard on getting enough muscle to pack it down, because it weighs 114 pounds. I got to where I could take it down there, set it up myself, dismount it, take it back up, and put it away. Then he told me they weren't going to teach me how to drill.

I said, "Why? Because I'm a woman?"

He said, "Yeah, because you could get hurt very easily. You got insides to get hurt that I don't have." See, when you're drilling, it will vibrate and sometimes they have to lean into the drills, and that much vibration on internal organs, especially female organs, is not good. He was smart enough to know that.

I said, "But then why the hell have I been packing that stupid drill up and down the God-damned drift."

He laughed and said, "Because I didn't want to."

He then took me up, and he let me push the plunger to blow a round of dynamite in the mine. Oh man, talk about a rush! I love to blow things up. You push that plunger down, and that dynamite blows. You can feel the vibration under your feet, and then you see that smoke start curling up out of those holes. It was great.

∞

We used to get off a swing shift about 11:00 p.m. A guy would guard the door while I took a shower. I'd get dressed and run down and open the doors to the furnace. Put my hair over it. I had long hair then—took only about ten or fifteen minutes to dry. I'm surprised I didn't burn it off.

In those days up at the Grand Imperial, they had a band six nights a week. Those miners worked hard. Worked all day and drank all night. Danced up a storm. Feet so damn sore the next day you couldn't hardly even walk on them to go to work, but you went anyway. Then you'd get off work, get cleaned up, and go out and do it all over again.

I dated a couple of those guys, but it doesn't pay to go out with people you work with. That way you see each other twenty-four hours a day, and that's not good. It helps when they're off and gone every once in a while. Send them on that hunting trip, send them on that fishing trip. Let them go boating, skeet shooting, play baseball, or whatever. Hell, yeah, pack them a lunch and help them out the door.

There were bunches of women here, it seemed like, but in reality there were like six men for every woman. A lot of those guys, though, didn't actually live in Silverton. They had families and so forth some place else. They were what you call tramp miners. They worked this job for a while, and then they'd up and go to another camp. Go home periodically. Resident miners stayed here, had families here, and never left unless they had to.

I've been bartending at the Miners Tavern since about 1997. I hate throwing my friends out when they get too drunk, but after the mines quit, a lot of the miners started drinking more than usual. It was like the whole town was a family, and mining was the family business. And when it went, it hurt a lot of people.

They couldn't find any mining jobs, because there were so many places closing. The areas where they were mining got overrun with help. Most of the old-timers that have retired here now, they get real sad when they talk about it.

There are those that will never get it out of their blood. They got it in their heart and their souls. They'll find a hole in the ground, if they have to dig it themselves.

[Pat Donnelly passed away in the Winter of 2001, after a hard-fought battle with breast cancer.]

Frank Gallegos

Bondad, Colorado

My hands are starting to hurt a lot. Arthritis, from twenty years in the mine and that cold, cold, cold. My dad moved to Silverton in 1941, just after I was born. He went to work at the Mayflower Mine. We followed. My mom and I, we rode the Galloping Goose, which was like a bus that ran on the railroad tracks. They would haul maybe twelve people at a time. We caught that from Dolores to Durango, and there we caught the train up to Silverton. We got there in 1942.

When they first started running the tourist train to Silverton, us kids would set up a stand. We used to go to all of the different mine dumps—my mom, my brother and me—and find these little rose crystals. Then we'd go by the railroad tracks, where they dumped that ore stuff. We'd get nice lead, iron, and pyrite. We'd sell all this to the tourists. That was one way we made money.

I also sold chipmunks to the tourists. They would pay five dollars for a chipmunk. Then they'd go and stick their finger in the cage. The chipmunk would bite the hell out of their finger, they'd drop the cage, and the chipmunk would run off. I'd catch it again and sell it again.

In winter, my mom would fix Dad a real heavy quilt out of all the old Levi's and stuff. Pad it real good. The miners used to ride the ore buckets two miles across the canyon, from the mill to the mine. Mom fixed him a little stool with a pillow on it, to put inside the bucket. He'd sit on that and cover his head with the blanket to keep warm. He said the wind would catch those buckets and just swing them like crazy. He used to tell stories about the wind blowing tram buckets off, with men in them, and they'd just drop into the canyon.

When they blasted in the mine, before they had the new drill machines, they used machines they called "widow makers." That was because they didn't have water, so there was dust, and the miners got to breathe in that dust. But Dad never complained.

They used to have a commissary up there, a store at the Mayflower Mine. One payday, my dad came home with a whole sack full of candy for us kids.

After they started shutting down the Mayflower, they called my dad to work up at Idarado. Like most of the kids who grew up in Silverton, when I got out of high school, in 1960, I also went to work at Idarado. I worked with my dad for eleven years.

They said that the dry drillings my dad breathed in all those years really messed up his kidneys. He ended up doing that dialysis for about eight years, but he was a stubborn man, like most miners. He got to where he couldn't even climb the ladders, couldn't even walk, and yet he kept mining. I watched him die in there. It took him eight years to do it. He finally died December 6, 1988. He wouldn't have had it any other way. All told, my dad worked underground thirty-nine years.

I went to work at Standard Metals. It was different at Standard. Standard had a whole different breed of man from any other mine around this area. We decided we didn't want a union, but we did form a grievance committee, one time, when we had a bunch of trouble. The company didn't want to give us any pay compared to what these other mines were giving. They thought because we weren't unionized, we couldn't force them to.

Well, we got over there to the main office, and all the bosses took off scared up the mountain, on these four-by-four deals—all terrain vehicles. They watched us down at the office, from up there. After a while it started getting hot, and all of us were getting a little tired of sitting there. The bosses wouldn't come back to talk to us.

A couple of us guys got together, and we said, "Let's take up a damn collection." We went down there to the liquor store, and we bought about $170 worth of beer. We got the guys there drinking beer in the hot sun. Got them madder. They weren't about to leave. Finally the bosses came down.

All of us got around those damn four-by-four deals, and we picked them up. Shook them a few times. We said, "We're going to talk to you." The bosses got out and went inside the office. Our grievance committee went in there too, and we got our new contracts.

That was one thing you didn't see at Idarado, or Revenue, or any of those other mines. Over there the bosses would tell them something, and they'd bend over and say, "Well go ahead and kick me again." At Standard we were united like you wouldn't believe.

There were 120 of us that were underground that Friday before Lake Emma caved in the mine in 1978. I guess the hole that went through there was the size of two football fields. They said on a rainy day, you could look down that hole and see the gold, like spider webs. And the guys went after it.

The company hired security guards, but even the security guards caught the gold fever. Some of them stole gold from the mill and put it in the back of a pickup truck; like the head security guard and two or three others. Authorities caught them clear down around Texas, where they were trying to sell some of it.

I tried sticking it in my shorts. That hurt. And then the gold situation in that mine got to where it was darn right dangerous. Guys were after it so bad that they took guns underground, and they'd pull them on each other. Threatened to burn each other out of the manways, using paper and boxes—anything to light and make smoke with. I figured, nope, I don't want no part of it. The gold wasn't worth my life.

The thing that I heard from the old-timers, when I first went to work up there was: it was a miner's right to take a specimen or two. The miners never really see but maybe one or two percent of all that gold that's taken out of there. And if you got a specimen, or you later on got a pie can full of it, it was considered a miner's right—as long as you didn't get too hoggish about it. And you didn't waste your shift high-grading. You still had to do your work, what you were hired to do.

My grandfather used to say, "Gold belongs to its owner. There's people that can walk by it and never see it. When the one it belongs to comes by, he'll find it." I always thought about that, when everybody started high-grading like crazy. If someone went in there to take it and sold it, that gold never did them any good. It never did me any good. I never prospered from it. It was money that came too damn easy, and there it went. It was gone, just like that. It happened with everybody. It didn't belong to them.

It took us two years to clean that mess up after Lake Emma came in. That mud was so dangerous. It was pure mud, and junk, and stuff that came from the bottom of that lake. If you ever got caught in it, that was it.

Years ago a friend of mine and I at Standard Metals, we were working at what they call the main level stope. It was all mined out, but there was a lot of mud. I was in there running a mucking machine. My partner was back there, running the muck train. I was loading it, and all of a sudden I could hear this strange slushy noise. I stopped the machine. I went back in the hole and got another bucket full of muck to dump in the train. I stopped there, half hesitating to go back in there again with that machine. All of of a sudden—woosh!—the three-ton mucking machine I was running, was gone. I turned around to run, and I fell alongside the first car I was loading. We had six cars: the five ore cars, the flat car, and the diesel motor in the back.

I fell there, and I looked back. I couldn't see the mucking machine. Then I looked up, and it was right above me. The mud had picked it clear up. Boy I'll tell you, I was crawling on my hands and knees. I got on my feet, and I ran. My partner says, "Shall we move the train out of here?"

I said, "To hell with the train. Just leave it and get the heck out of here!" So we turned around, and we started running. We just barely turned right and got behind the old shop down there, and that mud went right past us. It hit through the train, the cars, the mucking machine, everything clear almost down to the main line.

After we got going again at Standard, there were still places that we never went into, to try to clean out. It was all mined out—there was no point messing with it. But the thing that was dangerous about it was that there was a lot of that mud and stuff that would be sitting there, and all of a sudden it would turn loose.

My brother, Carl, and my friend, Tommy, were pulling muck that they dumped from the upper levels to the main level. It was spring and you get a lot of water run-off then. They were taking turns running the chute. There was one upper door in the chute for loading the muck into the chute. The lower door was to release the muck and load it into the cars. One guy would park the cars under the chute, and the other would load it. My brother was running the twenty-ton diesel when the mud came. He said all he could see of Tommy was his head going back. That muck caught him, and it just buried him.

I was working the upper level, at the time, and they called us and told us what happened. I knew Tommy and my brother were working there, so I beat ass right down to it. We started digging and all that kind of stuff. We pulled the cars out, because it buried the cars. My brother said, "I couldn't do nothing. I had to run." It took us about twelve hours to find Tom.

The bosses didn't want to go over and tell Tommy's wife he'd been killed, so I had to go and tell her what happened. I did it, because he was my friend. We grew up together. But that was the last time I ever did anything like that. After that I told the bosses, "That's your job. You do it."

The men were very close. Christmas time, we'd have some parties up there you wouldn't believe. The company would buy us a couple cases of whiskey. They used to give us a turkey, a box of goodies, and a bottle of wine. One time we were on night shift. We got up there to the dry room, and the day shift was still there, drunker than skunks. They were lying on the benches.

Later on we got to Silverton, went into a bar, and continued to party. Then we finally decided we were going to go home. I commuted from Durango. So did a lot of guys. We had about four carpools heading over that night.

I was driving, and all of a sudden that damn car started spinning. The guys in my car were saying, "What's going on?" But they were half passed out. We spun around two or three times, and it ended up going back the same way as Durango, so I just kept on going.

We got up to the top of Coal Bank Pass, and some of the other cars were already there. Piss stop. We then started to play football with our turkeys. I don't know how those turkeys survived, and I don't know if they wound up being very good to eat, but that was fun. We used to have a lot of parties like that. The guys were very close. We had a lot of characters.

One time I was working with Billy Hunt. He was an expert. I let him down the hole with a rope. He had fifteen cases of powder. He put a primer on each end. He lit one, and he was over there with a cigarette. I told him "Billy, God damn it, get the hell out of there. You lit one. Light the other, and let's get the hell out of here!" Fifteen cases of powder would have smeared us from here to "kingdom come."

Frank Gallegos (right) working with his father, Max.

He said, "I-I'll be right there." As soon as he grabbed that rope—shoom!—I pulled him out. There was a cement deal we would get behind, when the blast went off. I tried to get billy to hurry. Finally I picked him up, but I didn't get him behind there by the time the blast went off. His cigarette was all kind of bent.

I said "You all right, Billy?"

He said, "I guess. Let's go eat lunch." That was Billy for you.

I got sick in August 1983, to where I couldn't take the altitude anymore. We were going up about twelve thousand feet above sea level. I couldn't breathe.

I went to the doctors, but you know, being hard-headed like most miners, I went back to work. Well, I tried to, but I couldn't do the work. They kept sending me out. The third time I went back in there, the safety engineer and one of the main bosses called me in to the office and told me, "Frank, that's it." At the time I left, I was the miner that had worked there the longest.

There is going to come a day when they're going to need these metals again. And they're going to want somebody to mine them. Well, there won't be no miners around with any experience then.

Maybe they can buy metals cheaper from overseas, but what are they doing to the men? That was an uneducated man's job. It was an honest job. A lot of the men didn't even graduate from high school, but they went in there, they made top dollar, and supported their families. Now those men are all gone. Their kids are going into different jobs.

Truth is, there'll be no miners around when they need them.

Joe

["Joe" is a fictitious name. He wished to remain anonymous,
due to his accounts of high-grading.]

When I was a kid, we used to sell ore on the street corner to tourists. All the kids in Silverton did it. The miners would bring rocks home, and the kids would take them out and sell them. The rocks weren't usually gold specimens. But some of the kids, who were taking rocks without their fathers knowing it, had gold in their boxes and didn't even know it. Shop owners would look in the boxes and find chunks in there that were worth $30, which they could buy for $1.50.

Later when I was working at Standard Metals and high-grading, guys from the Mormon church used to come with brief cases full of cash. Open them up and give you hundred dollar bills for your gold, which we'd weigh on the bathroom scale. Probably not very accurate. I was kind of freaked out when they first came to my house. It was a couple of guys in suits. Right away I thought it was the IRS or the FBI. But that didn't happen until later.

High-grading was wrong. If you steal a nickel from someone, you're stealing something from someone. But I guess it was our way of getting compensation for as dangerous as it was. Or maybe that was just the excuse. Everybody in town was high-grading, except maybe the preacher. And pretty soon we had him going too. The preacher was also the hoist man. He hoisted about twenty men at a time on this elevator, between the inside of the mine and the upper level, which was about a thousand feet up. We'd take a couple chunks of high-grade out of our lunch box and leave it by the hoist. Next day, it wasn't there.

When I was first married, I think my wife knew I was getting a little bit of gold. I had a lot of money. I bought a new car. But she didn't really know that much about it. Our shifts went from 4:00 in the after-noon until 2:00 in the morning. By the time I got home, she was asleep. Well one time I was so tired, I took my gold out of my lunch bucket and set it on the counter, then went to sleep. When I got up, I went to the counter, and the gold was gone. I said, "Where's that stuff that I had on the counter?"

She said, "Oh those rocks! I threw them out on the street. And don't you ever leave dirty rocks on my counter again!"

So I went out in my shorts. Tourists were watching me looking for my pieces of rock. I brought the gold back in, I washed it, and I sold it that afternoon for about $600 bucks. My wife watched this whole thing go down, and her eyes were big. The next night she was waiting up for me with a scale. She started talking about a new refrigerator and a new dryer.

There used to be a back way into Standard Metals. Nobody was supposed to be there. That's how we would sneak in on the weekends. We'd turn our lights off and hide from each other, kind of like cat and mouse. You don't know who's standing over there in the dark. You might tell by their voice, but it was always better not to confront someone under there, when nobody's supposed to be there. It was pretty crazy.

The bosses were usually the first ones up there high-grading. They got the pick of the choice rock, and then we'd show up. They threatened me. They told me, "We don't want you up here high-grading. If we catch you tonight, we're gonna fire you." But in the end, they probably figured what's good for one is good for all.

Some people brought out a lot of high-grade on the weekend. One weekend I went underground, and they had the big compressor going downstairs. People were drilling. I thought it was a Monday or something. They were just going at it. Probably were bosses and other friends. You know, "Let's go get some gold and make a $100,000 this weekend." They'd go in there together and just rock 'n roll.

But the bosses sent you to work, and if you weren't in your working spot they could fire you. So me and a friend of mine—he was my best partner—we worked it so we were trammers, hauling the ore out and dumping it down the chute to the main level. And we were still making about as much as the miners, because one of us would do the tramming, and the other one would get gold all day. Toward the end of the day, we'd bring the gold down. So we were making more money than most people thought.

One time I was up in a stope getting some gold. A stope is like a big room where they're drilling and stuff. They had hit some pretty good gold up there, and it was just piled everywhere.

My cousin was there too. He was pretty greedy. He kept stacking up gold. He probably had it six feet high, seven feet wide. He just kept throwing rocks onto his stack. I went and got some boxes and started boxing up my stuff. And that guy I was tramming with, my partner, he came up and helped me take it down.

Well, it was getting close to quitting time, and my cousin just couldn't stop getting rocks. He kind of screwed himself, because he piled it all up, but he never went and got any boxes. So when it came time to move it, he wanted me to help him go get some boxes. I said, "You know this ain't the way it works. Everybody's on their own. I need to get my stuff down and get out of here in time to catch the main train down below, or else they're going to start asking where we're at."

He looked around and didn't know what to do. So he took off his diggers, which is like a rubber suit— rubber coveralls. He tied the bottoms of the legs and arms, and stuck the gold in there. He probably had about 150 pounds in this rubber suit, and he was dragging this thing, and it's stretching and shit. He actually got it down to the manway, where you walk up these ladders to get out. He was climbing up the ladder, pulling on this rubber thing with its arms and legs full.

I had left before he did that, and I was going back to get more of my boxes. When I came back, I looked up there and here's this thing that looks like it's seven feet tall. It freaked me out. There's this thing we call a "stope ape." Guys are always talking to you about an ape being in there, or some man that's made out of rock or whatever, and it's going to grab you. Well, when I saw this thing, I lost my breath because it looked like a stope ape to me. This sucker had arms and legs. And my cousin's pulling on it, so it's moving. But then it ripped. And I just said, "Ah, shit." I don't think he talked to me for six months, he was so pissed. He blamed me for losing his gold.

∞

I think the mine owners knew what was going on. One time I went to a meeting down at the Standard Metals office. I walked in there, and the boss said, "So what are you bitching about now?"

I said, "Well, we have some grievances. We want to talk to you about these certain things."

He said, "Pick that canister up." They had this canister of gold. It was pretty heavy. I set it back down. He said, "You know how much is in there?"

I said, "No."

He said, "About $180,000 worth of gold."

I said, "Wow!"

He said, "That's where I get my gold. And I know where you get yours. So I don't exactly know why you're here. I don't think you have any business complaining about anything."

One time I got fired from up there, because I was accused of running off with a semi truck full of gold. I was the watchman outside the back way to the mine that weekend. Supposedly I had some friends come up, and we went underground and took a whole semi truck of gold out through the back way. That's how the story goes, even as of today, but they never proved anything. Now when I go up to Silverton, these old-timers buy me a drink and say, "So how about that story—what's the truth?"

I say, "Well if you buy me another drink, I might tell you." It's a good way to get drinks, but I'll never tell.

That rumor got out about me and the semi-truck, and I had the IRS and the FBI come to my house, to talk to me about my high-grading. There was a black guy and a white guy. They stuck me between them in their car, and they turned on one of those tape recorders. My father was looking out the window. He thought they were going to haul me off. They asked me questions, like how much gold I'd taken this year and that year. What was my biggest find? Who else did it? Was everybody doing it? Were the bosses doing it?

I just kind of acted dumb toward a lot of the questions. Then I said, "I got a question for you: If you found that gold, would you take it, or would you not take it?"

This guy said, "I can't truthfully answer that."

I said, "Well, I can't truthfully answer any more of your questions. You're telling me you might do the same thing I would do, and you're over here questioning me about it." I said, "I'd like you to turn that tape recorder off, and I want out."

They opened the door and let me go. That was the end of it. They just rolled away. Maybe there was no evidence, or maybe they wanted me to be a snitch.

A lot of people were freaked out about me for a little while, because they were worried I had snitched. They confronted me about it, and I said, "Don't worry." The FBI never came back to bother anybody.

Some people got caught high-grading. One kid came out and had a bunch in his boots. They were doing one of their random searches. He couldn't back up, and he didn't want to go forward. He was kind of stuck. He got to the point where he had to go through the check, then when he got there he just started running. He kicked his boots off, and he started throwing the gold rocks all over the place. He got away with it, somehow. I don't remember the technicalities, but they sure scared the crap out of him.

There's also been other people who have been "talked to." A friend of mine had a lot of gold in his trailer. He went home after our shift, and someone had ripped the door right off his trailer. They'd taken all his gold, a bunch of his rifles, and other stuff. So he called me up. He was all drunk and pissed off. He drank about a half a quart of whiskey, bummed out because he'd lost a lot. He said, "Hey, you need to come over, man."

I said, "What's going on?"

He said, "I'm sitting in my shorts, looking out the door, because there is no door. And someone took all my gold."

I said, "You didn't call the cops did you?"

He said, "Yeah, I did."

I said, "How stupid can you be?"

I think it was actually written up in the paper: "SOMEONE STOLE MY GOLD!" I mean—come on now. The obvious question was: "And where did you get the gold?"

There was another friend of mine, a kind of goofy kid. I was over at his house. He took some gold that he had and stuck it in the bathtub. He jumped in there with the rocks and started washing it up. He said, "You ever tried this?"

I said, "You're sick, man. I think you gotta get out of this environment."

The next week he was on his way to Denver, and he had some gold in his van. He was messing with a piece, as he was driving down the interstate, and it fell between his feet. So he stuck his head between the steering wheel, trying to reach for it, and couldn't get his head back out. He had to stop in the middle of the interstate. They wound up having to cut the steering wheel off his head.

If I had the time, I'd like to try prospecting. I can smell gold. When I worked at Standard Metals, they told me they wanted me on the lower levels; because the upper levels had the gold, and they wanted me away from the gold. So they put me on G level. Well, this rock, that was about as big as a table, came out of the chute and landed in the tramcar. I was going to use dynamite on it, then I saw these little specks of gold. So I just started breaking it up with a sledge hammer.

Then here come all the bosses, and they said, "We don't know what to do with you. We don't know where to put you to keep you out of the gold."

I said, "Dude, this thing just fell out of the chute. What could I do?" They just shook their heads and walked away. I got about three boxes out of it.

My biggest find was a rock about the size of a Volkswagen bus. It rolled out of an ore pocket I was slushing and landed on my slusher bucket, which pulls the ore and dumps it down into a chute. I broke my cables trying to get the bucket out, so I took some dynamite and blasted the rock. It was loaded with gold. Big ore veins all the way around the whole thing. I went crazy. I took some chunks and hid them. Two nights later, I brought it out—ten boxes—and sold it for $15,000.

I went back to work up at Standard a few years after I'd been fired for high-grading. I don't know why they hired me back. I got a job loading trucks with gold.

I decided one night that I would get a truckload—not a semi, now. I had an old antique pickup, and I took it over there and loaded it up with about a ton of ore. I was bringing it down to my property and I blew a tire. I was on the side of the road when the bus and all the bosses just drove right past me. I got lucky. As for the ore, I found very little gold in it, so I made it into a rock garden. That was the last time I would attempt to do any gold craziness.

I could never understand why gold is worth what it's worth. This friend of mine always says, "If I had to choose between gold or coal, I'd rather have coal, because at least I could keep myself warm." He says, "When minerals aren't worth anything anymore, I'd rather have a pile of coal sitting outside than a pile of gold." I guess that's why we all got rid of it. One of my buddies would give his dentist gold for working on his teeth and also get his gold fillings done that way.

I think it was just the game. It's kind of like gambling. Why not try it. Get some excitement in your life. My biggest regret is that I never got smart with my money, investing it like I should have. I was too young and partying too much. I always thought there was going to be more tomorrow. Nobody thought Lake Emma was going to cave-in the mine.

There were people that went up to Silverton, new miners, and they'd say the reason the mines were closing was because so much gold was stolen; but we couldn't have even begun to touch how much money was made in there. And here's some guy, who's just starting up there, and he's being loud and saying, "This place would have made damn better money if everybody wasn't high-grading."

Guys like that usually got their asses kicked and run out of town.

Leonard Williams

Norwood, Colorado

My dad was a farmer. He got bucked off of a horse when he was young, and he hurt his leg. Limped all the time. Eventually he had to have his leg taken off. He was a super good dad, though. Made a living on the weed patch. He took me everywhere he went. I suppose we were poor, but my mother was a really good cook. We were always clean and had a full belly, so I thought we were rich.

My brother mined a little bit, but he said he'd be underground long enough when he was dead, so he went to work in the oil fields. I was about eighteen years old when I started in the uranium mines. That was about 1959. I had kind of a tough time getting on at first. I looked like I was about twelve.

After a couple of years, I went to Telluride. I guess I shouldn't say that I liked it underground, because you get powder headaches and so on; but you made good money, and I didn't have enough sense to be afraid of anything.

My friend, Buckwheat, got hurt pretty bad one time. I was right there when that slab fell on him. I don't know how he survived it. The slab just kind of bridged over him on other things. We got him to a safe place as fast as we could, and he made it.

There were more accidents on that muck train than anywhere else. Albert Moreaux Jr. got killed on the muck train. He graduated from high school a year ahead of me. He was in his early thirties when he got pushed between that car and the chute.

We heard all about it from George Cappis. George hauled tons and tons of ore out of there on the muck train. He was one of the orneriest people who ever lived, but if George didn't know something, he kept his mouth shut. What he did say, you could take it to the bank.

Those guys I worked with, they were characters! Aubrey Lillard, "Old Blizzard," drove miles of drift. At one point he couldn't pass the physical to go to work in a coal mine, so he got on state comp. He said, "It's gonna cost them a lot of money, because I'm gonna live to be an old son of a gun." He always had an answer, if you got him wound up.

Buckwheat used to say, "He's the only one who could go to town with five dollars and stay drunk for a week." Buckwheat was always wound up. I wouldn't say he defied authority, but he'd catch the biggest boss in there and tell him what he had on his mind. It was humorous. Buckwheat didn't give up easily.

Then there was Fred. He always had a story to tell. He's been gone quite a while now—cancer. The doctor told him, "You're going to have to have surgery."

He said, "If I don't have surgery, how long will I last?"

"Oh, maybe six months."

"If I have surgery, how long will I make it?"

The doctor said, "Maybe a couple of years."

Fred said, "Well, six months ought to be enough to get everything done I have to do." He lived at least ten more years.

It was a big world under there. The Idarado goes from one end of the mountain and comes out through the other. We had stopes that averaged thirty-five to forty feet wide. That was opening up a lot of ground. You'd kind of farm it, and there were a lot of big boulders.

The upper levels were pretty cold. The lower levels were pretty hot. There was a lot of difference. You'd kind of dress for where you were going to be working. If something happened, and you got sent from the warmer lower levels to the upper levels, you would get real cold. There was a dog house up there, where you ate lunch. It had an electric heater and an electric light. They'd box it in with two-by-twelves on a couple of sides and have it enclosed, so it would be a little warmer in there.

We used to give the crew that operated the muck train a quart of whiskey every pay day if they'd keep our chute empty. When our chute stayed filled up, and the ore wasn't moving out of there, it was costing us money. One time we were kind of down, it was a slow time. We got our nipper a quart of wine and told him, "Well, that's all you earned, you didn't do anything." He gave us a bad time, so then we got him his whiskey. He was Albert Moreaux Jr., the one that got killed on the muck train.

Some places the ground would get pretty rotten and treacherous. You'd get nervous, but you didn't let it bother you very much. You just did what you had to do. I remember Buckwheat and I were barring down a slab in a finger, about two rounds up. We were standing on one and barred a slab down on the other one. It broke it off about six inches from the end, close to where we were standing. That takes a pretty good size rock. You just kinda chuckle and say, "Glad we weren't under that one."

I guess I've been accused of dodging a few slabs. I had a lot of close calls. If you're young and foolish, you don't think much of it, but I had nightmares about slabs. My wife would ask me, "What were you dreaming about?" I guess I was taking off in my sleep, to places and situations I didn't want to go into during the day.

I did actually have a slab come down on me once, but the end that came down on me was a little bit thin and talcy, so it broke over the machine. I wasn't really hurt, yet it did monkey my shoulder up a bit. I was put on light duty for a couple of weeks.

I worked in the Idarado probably thirteen years. It was a good place to work. We lived up there, in Telluride, maybe two years. It was a real friendly town. I knew eighty percent of the people. If you didn't get to the basketball game early, you didn't get a seat.

But then we bought a couple of acres near Norwood. We had horses, and there was more interest down here. Plus, in Telluride, there wasn't anything to do in your spare time except go to the bar, and I didn't feel very good after I did that.

So I commuted from Norwood to the mine on a bus that Idarado ran down here. I used to drive the bus on night shift. One time, going to work, we had a big flash flood. There were boulders. We didn't think we would make it to work. We just kind of piddled along, sort of like sightseeing, but we made our shift.

I have to say I did like mining. I didn't like the conditions sometimes. If you were working in a gassy place or had to eat a lot of powder smoke, stuff like that, you didn't enjoy that very much; but I would have never been able to buy my place if I hadn't worked there.

I was on a crew with Buckwheat and Don, Cecil, Hank, and Roy were on the other shift. We made top money for several years there. When Roy died, Les came with us. We'd give each other a pretty hard time, but instead of arguing and complaining with each other, we helped one another.

Miners are kind of a sinning sort. They did their girl chasing in the mine. They did their hunting in the mine. And they did their mining at the bar. A lot of them drank way too much. Most of them smoked too much, and it was pretty hard on their health. But when you were in trouble, you had a friend. They'd give you the shirt off their back. They were good, hard-working people.

I have a soft spot in my heart for those old miners.

Bill Young

Olathe, Colorado

I started at the Silver Wing Mine in 1959. I was probably about eighteen years old. They called me "Swiney," because I had a pig valve in my heart. When I first started really mining, underground, it was up at a uranium mine at the foot of Monarch Pass near Salida. The three years I spent there, I did stope mining, drift mining, a lot of timbering. In a short period of time, I probably did every facet of mining that could be done. Then I got caved in on and broke my neck. I was off for about a year.

I went back to mining. I worked at the Idarado thirteen times. I would get pissed off and quit, and then I'd go back. As long as you could drive drift, they'd hire you; but I never could get along with Idarado, with all their safety bullshit. I don't need all that. I figured I could take care of it myself, and I still figure it that way. I mean that's what shut that mine down, all that safety shit. The government's going to save everybody's lives? That's bullshit.

All the accidents I've had, I never felt fear. Generally when I get hurt, I get mad. First thing I do is I blow my cork.

One time I got caught underground. A transformer was burned up. Electrical smoke. I came out of it, me and another guy. But the third guy, Jerry, didn't make it out, so I went back to get him. He was floundering around in there. I told him to just grab the pipe and follow it out, because you couldn't see anything, so he made it out. But by going back to get him, I just took on too much smoke. Got my heart enlarged half the size of normal. It took about six months for it to go down. It really affected my health.

I liked mining, at first, making all that big money, but I outgrew that real quick. It's pretty much always been a job to me, and the money isn't any good anymore. I've been at the Grizzly Bear Mine since about 1990. The Grizzly Bear is where Joe Mattivi just got killed.

Nothing nostalgic about being a miner. You got to wear a size fifty-two coat and a number two hat to be a miner. Anybody missing mining has got to be an idiot. I only miss going to the restaurant up in Silverton and teasing all the little waitresses. But nobody should miss mining. I hate it. It's a dirty, scroungy, lousy, miserable job.

There's a lot of things I like better than mining. I like anything better than mining. I'd go to the bars and talk about cows or sheep. I talked about women, but I never went to the bars and mined. When the sub-

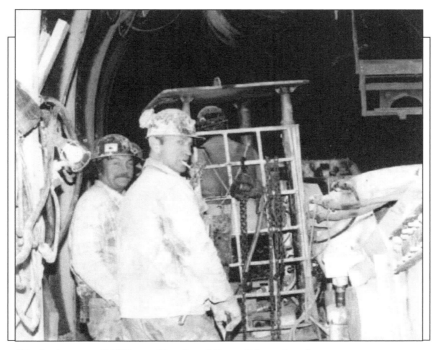

Hardrock miners Bill Young (front) and Leroy Palmer (rear).

ject turned to mining, I got up and left; and I never did bring my mining home with me.

If you really want to get down to it, most miners are old drunks. I don't know why. Why are painters always drunks? Why are sheet rockers always drunks? Damned if I know. I work hard. I play hard. Everything I do is hard.

I watched this movie, one time, called "Lonesome Dove." Old Gus said, "Life is short. Shorter for some than others, and that's the way it is." I always kind of remembered that. Life is short. You get old. You don't even know you're getting old, until you're there.

Ron Williams

Montrose, Colorado
Interviewed at Hardrockers Holidays in Silverton, Colorado

*A*fter I graduated high school, I got on with the Forest Service over in Montana. During the winter, when I'd get laid off, I would go to see Grandma, Mom, and Dad in Delta, Colorado. One winter the girl I was going with, her friend's boyfriend worked at the Camp Bird Mine. He said, "Well come on up and rustle."

I said, "What do you rustle?"

"Hell, you rustle a job."

So I went and rustled a job up there. I started at the Camp Bird on January 28, 1961. I was nineteen years old. My first time in, I was sitting next to the machine doctor. He must have been laughing all the way in, because I kept ducking, and I must have had four feet of clearance.

My shift boss's nickname was "Whispering Jim." In a normal voice, you could hear him clear over to where the road was coming in. He stuck me up on the ore pass, just keeping it clean. They would dump the ore, then I'd get out and break the rock, so it could go on through.

When I first went in there, in the ore pass, for some reason my light started going out. I couldn't get another light, because we were almost three miles from the portal. Well, I knew enough to turn my light off and sit there in the dark. When they dumped, I'd hear it coming, and I'd just sit there until they finished dumping. Then I'd turn my light on, and I had enough time to beat the rocks through before my light went out again.

I worked at the Camp Bird a little over two years, the first time. I also worked at the Idarado and Standard Metals. I would take a couple weeks off in the summer for the National Guard. It got me out of the mines. That was my vacation, more or less.

I was always pretty serious when I went to work. This ol' boy I partnered up with one time, we hardly ever said a word to each other because we knew what the other was going to do. And when we went to work, we went to make money. It was Rod MacLennan. He broke me in. One time, he and I were drilling a V-cut. It's nice to have your two holes bottom out right together. He was looking down there, and I blew my hole, and that stuff came shooting out and caught him right in the face. Oh he was mad. Not at me. Just mad at himself for looking in the hole, I guess.

The Red Mountain area, it's got some heavy, bad ground. It was hard to hold. One time my brother and I were driving a raise at the Camp Bird, and we hit this water course. It had all that deposit of water, from over millions of years. Kind of muddy stuff. It was about four feet wide, and it was over one hundred feet long. It had that muck-hole sound to it. Your voice wouldn't carry at all. It would kind of just die out. That was a spooky feeling.

We had to widen it out to get through it. My brother and I, we went in and bombed the slabs that were hanging up there. We'd take a twelve-foot ladder with us. We would climb up, put dynamite behind a slab, and wire it all in. Then we'd go out to the blasting box, throw the switch, and shoot it. It was kind of a "picky poke" operation, but we got through it and nobody got hurt. That was the main thing.

My friend, Tommy, got killed in Standard Metals, on the main ore pass. He was on the tram. It was real wet and soupy. Then the whole thing moved at once. It flat pushed everything right out. He got buried under all that muck. Buried the tram and everything.

Ron Williams, August 1978, re-opening a mine tunnel.

When I went to work up at Standard Metals, the county called in, one time, and said, "Get the guys out. We're going to shoot the avalanches down." So we changed clothes, went to town, and had a beer.

There's about twenty-eight slides between Silverton and Ouray. They usually slide down the same places. In places where they had trouble, the old-timers would go up above the portal of the mine and build these big rock V's to split the snow, so it would go around. But once in a while you get one that lets loose up high, and comes down, and tears out a new path.

When we started back, it was Sam's turn to drive. We were heading up, and another slide started to come down. We saw it coming, but before we could get stopped, it hit the road right in front of us. We went—

poof!— right on top of it. Couldn't get off. We had to wait for the state to come up with the Cat and pull us off. We went back to town, and went to the bar, and had a few more beers. Then we headed on home.

∞

I had some close calls in the mines, but it never made me want to quit mining. I was working with one guy who collared some holes for me, and he said, "Well you go on and drill. I'm gonna go eat." He sent the nipper down with me, and it's a good thing he did. That drill had me pinned up against the back, and it was still running. I was pinned in such a way that I wouldn't have been able to turn it off myself.

A good friend of mine was killed, and he was a careful miner. A slab just fell on him. It was his time to go, I guess. But to get crushed by a rock was a hell of a way to go. It was kind of spooky that we had to set up a raise in a stope right next to that one he got killed in.

I guess the scariest experience I had was when I was sinking shaft. We had sent the muck bucket up, and I knew the mine mechanic was coming down in it. All of a sudden I heard a big banging going on up there. One of the two eighteen-inch guide shoes, which keep the bucket from swinging, fell to the top deck where I was standing. I jumped behind a rope. The next guide shoe hit right in front of me. The bucket started swinging, and it hit the steel and jammed in.

The walking boss said, "Well, somebody's gotta go up and find out what happened."

I was the lead miner, so I said, "I'll go up the rope." The rope had that real heavy dope on it—grease. I climbed thirty feet of rope, and it took all the strength I had to pull myself up on the landing. It turned out the mechanic had been in the bucket when it jammed, and something inside the bucket had smashed him in the nose—just peeled his nose back. He had blood all over his face.

So here was the bucket jammed in. The crosshead was hanging kind of cockeyed. I had to cut the screen to get out of the manway and over to him, but he was trying to get out and over to where I was at. He wouldn't stay put like I told him. I was afraid he would be in shock and fall, but by the time I got the screen opened up, he got over and got on the landing. He never missed a shift. We set the shaft sinking record, and I got some reward patches for it.

I eventually quit working there. I now work for a mine up in Nevada. The only thing that would make me want to stop mining would be my age.

Mining gets in your blood. Whenever I look at mountains, I think, "Is there any ore in those?" I'm always looking at the structure of it. "Where would the vein be?"

A "skip"—an underground elevator.

Louis Valdez

Durango, Colorado

When I got out of high school in Ignacio, I moved around looking for a job. I wasn't looking for a mining job, necessarily, but I wound up in Silverton. I started at the Standard Metals Mine in 1963. Once I got there, I knew that I didn't want to leave. Called it home. I stayed there thirty-one years. There were some good old days in Silverton.

There were a lot of people in Silverton at that time. We had about four mines operating, including Idarado and Standard Metals, and then some little independent mines. The bars were always full. All the women used to go with their men to the bars. Then after the bars closed down, we would all go party someplace else, somebody's house. At Christmas you would start at one end of town, and you just went from house to house. Everybody had some food on the table and hot drinks. It was one giant family.

There were a lot of Italians there, just coming out of the Marines. A few Germans. We even had one of Hitler's former ace pilots working in the mine. He was a tough, bad old guy. Al was his name. He worked by himself. How he got there, I have no idea; but he worked for us many years.

Everybody got along fine. When I first went up to Silverton, there were fourteen bars in that town, and you hardly ever saw a fight. If some outsider came in, then you might see a fight. You'd see a strange person in there; and, I don't know, that's just the way it was. When they started the Old Hundred Mine, they brought in a crew to build the sheds up there. We couldn't get along with those builders. They had an attitude. They also made a lot more money than we did. It turned into a fight. There was a jail at the town hall, but it was condemned, so they couldn't put anybody in it.

Mining was hard work, and the money wasn't very good at that time. I started at $1.45 an hour. The contract miners were making seven or eight dollars an hour, which was good money in the sixties; but I worked there about five years before I started contract mining.

The thing about Standard Metals was you never had to get up in the morning, back your car up, and go to work. You just put your coat on, grabbed your lunch pail, walked across the street, and got on the bus. The buses would come by and take you to work.

In the wintertime you hardly ever saw cars in the streets. Everybody put their cars in the garage for the winter. The only time they would take them out is if they had to go to Durango to get groceries. Other than that, everybody walked wherever they were going.

We missed only one week of work due to avalanches. That was when a big storm hit in the 1970s. Every slide in the valley that was up there came down. Red Mountain was closed. They got groceries into the stores with helicopters. A watchman was stranded up at the mine, but we always kept provisions up there in case that ever happened.

The Reverend and his daughters got killed in a slide on Red Mountain. I had gone hunting with him that year. I spent three days up there probing in the snow, trying to find the bodies. We found him and, a couple of days later, we found one of his daughters. The youngest daughter, we didn't find her until the spring.

Tony Jiron, Jr., Louis and Eddie Sandoval, and Bill Rich in the "dog house." Courtesy Louis Valdez

After a while I became a boss at Standard Metals. I started dropping the younger guys in with the older guys. The older guys kind of resented that at first, but then they finally realized that a young guy with a strong back has a lot to offer. I treated my men good. They were my friends, and they still are.

It was my job to go around checking to see how the men were doing, what they needed, supplies and all that. But I didn't know what they did after I left. They could have done a lot of things they shouldn't have done, which they probably did. They had a lot of shortcuts to do this, and shortcuts to do that. It's just the way it was. I know a lot of the miners, the old miners, they did a lot of things after the boss left.

We had a lot of trouble with those real young guys who came in and had a lot of schooling. The mine would put them in "bossing" right away, and they didn't know the first thing about how to deal with the miners. The miners knew what they were doing. They were sharp men, and they were skillful workers. The young, schooled guys would come in and say we had to do it this way and that way. They didn't understand how the mine operated. We eventually went to on-the-job training. We'd start them from scratch, just like anybody else.

A lot of people thought the mines were real dangerous. To a point, they were. But it was up to the individuals. You had to take care. Just like driving down the road, you could have a wreck if you didn't steer your car. Same way in the mines. You controlled your ground. You see those slabs hanging, you get them down, or they're going to fall on you. If you saw a bad area, you got ahold of your supervisor and said, "We think this area needs to be bolted and timbered. Do it."

I don't think regulations really made mining safer. In terms of health for the men, ventilation and such, it did help. Inspectors came in, and they made it so you got fresh air. But as far as ground control and stuff like that, I think it probably stayed pretty much the same. The miners knew what they had to do.

But there was about seven or eight guys I know who got killed in Standard Metals. One was a real good friend of mine. Me and him had started working on the same day. He fell in an ore pass. We used to have boards that we put over the pockets. We'd double them up. He was going across, carrying a box of powder, and the boards gave out on him. He went down. He was only about twenty-five years old. I had to carry him out. We put him in a little tool sack. That was all that was left of him.

The last guy I brought out of there was a guy we used to call "Cowboy." He was a sub-contractor. They were doing some mining up there by the old Mayflower. They were starting the drifting and weren't in there very far when they got into some bad ground. It caved in and smashed him.

I had just gotten out of the hospital. I'd been in for about thirty days, because I had pneumonia real bad. When I went back to work at the mine, I was still packing around oxygen. They wouldn't let me go underground, so I worked on the safety crew outside. When that happened to Cowboy, I was one of the guys that went in. We had to muck him up by remote control, because the ground was so bad in there. You kept thinking, "It could have been me." But you just kind of had to suck it up and go about your business.

I was the last guy to check the mine before Lake Emma caved in June 1978. We knew there was a fault up there, but we didn't know that we were that close to it. For a month or so, we were monitoring it quite a bit. We'd go up there twice a day and check it, see if anything was happening. That Saturday I had probably one of the biggest crews working up there, because we were short on supplies, and I had said, "Give me some men." I went up there twice to check it, and nothing had changed on me.

That Sunday night I was eating dinner at the hotel. It was raining really hard. Somebody came in and said the mine had caved. There was just a rumble—everybody talking about it. I couldn't get ahold of our watchman, so I told Hal, the mine foreman, "I think we better check to see what the heck's going on."

We got the company truck and started up to the mine. When we went over the bridge, it looked like tar coming down the river. We got about three-quarters of the way up there, and we couldn't make it any farther.

The next morning we went back, and the water had receded enough that we could get up to the mine. Water was still shooting out of a portal. We found the night watchman, and he was scared to death. He said that, initially, the power had gone off. He had walked from the dry room to the power panel, near the portal. After throwing the switch back on, he started walking away. He said, "I could hear this rumble. It sounded like an airplane over the top of me. All of a sudden that whole side just blew—boom!—and water came flying out." He was still shaking that morning.

The lake was empty. I was just thankful it didn't happen when we were in there, because that was the biggest crew we'd ever had in there. It would have killed them all.

∞

My oldest daughter, if she could have gone into the mines, she would have. She's a tough girl. She likes challenges like that.

When I first started, if there would have been a woman working in there, the guys would have walked out. They thought it was bad luck to have a woman underground, but they kind of came to accept the fact that women were allowed to work under there. A lot of the women who went to work for me, they were good workers.

But I told my daughter I wanted her to go to school. Mining was a good living. It was good money, and it gets into your blood, but I didn't want my kids to work in the mines. I felt that they ought to pursue something else, something that had more of a future.

Two months ago I was offered a job mining in Venezuela. I said, "No, I don't need that." I can't even raise my arm up, because my shoulder doesn't work anymore.

Rick Ernst and Roy Andrean

Interviewed during the Hardrockers Holidays in
Silverton, Colorado
Rick lives in Dove Creek, Colorado
Roy lives in Chandler, Arizona

Rick Ernst (left) and Roy Andrean (right).

RICK: Roy and I were best friends. We went to high school together. All of us started working up in the mine during the summers. Just about every one of my classmates went into mining.

I went to college for two years. Came home summers to work in the mine and make money to go back to school. First I just ran the slusher and did some tramming. Then I got into mining. At that time we were making really good money. I figured why go through two more years of college when I liked mining.

ROY: My dad had come here from Sweden in 1931 to mine. He met my mom when she was a cook over at the Camp Bird Mine, out of Ouray. He worked in all the mines here. In the boom days, in the forties and fifties and sixties, there were probably twenty to thirty active mines going, so he just tramped around.

RICK: His dad broke in a miner named Gene Miller, and Gene broke me in. I had started at the Standard Metals Mine, as a nipper. I was a pretty big guy, so they wanted me to get to mining. I went to slushing and tramming, and then finally Gene Miller took me on. He became my mining partner for years.

You looked up to these guys. Some miners have a reputation you just admire. There's bad miners, there's good miners. Good miners get a lot of respect, because they're the ones that are producing. They're the ones that are breaking the rock.

ROY: My dad died when I was nine. From silicosis.

RICK: My dad was a miner too. He's on oxygen support now.

ROY: The guy who I looked up to most was Billy Rhoades. Silverton was a gold mining town. You didn't want to be one of the slouches. So there were way more good miners than there were bad miners, because if you didn't work hard, then you didn't stay a miner. If you weren't good, then you'd go back to being a nipper, or a slusher, or whatever.

RICK: A lot of times they would put younger guys with experienced miners, drillers who can teach the young guys and bring them along, because the faster they can be taught, it just benefited the company.

When you got with a partner, and you worked good together; it was years that you worked together. They wouldn't move you once you got experience—"Okay, now you train somebody else." It didn't work that way. You got to be partners for years. There'd be sometimes your partner would miss work, and they'd give you some other guy, and there's some people you just don't get along with.

ROY: The main thing was the safety factor. You have to work with this person, and it's like a second wife, basically. Damn well better get along— you better like this person, because your life is in his hands. I mean you're handling powder, and blasting every day, and taking care of the area you're working in—making it safe, or trying to.

RICK: If you take shortcuts in mining, you might not be around tomorrow. Yeah, you might only have a broken leg or a broken arm, but when you're off work, you're not getting your money. So you had to take care of yourself, and you had to take care of your partner.

As far as accidents go, I only had a broken foot once, but I didn't miss any work. I had a cast on it and wore a size thirteen overshoe. Didn't miss any work. I dislocated my knee and stuff, but nothing major. My wrist, but that wasn't really from an accident. Maybe just wear and tear. Just getting kind of wore out, I guess you'd call it.

ROY: People got a lot more sick in the old days. They used to drill dry. All that stuff would get into their lungs and cause silicosis. My mom didn't think it was too neat that I went into the mining industry, but you know, you're eighteen years old, and you think you know it all.

Now they have it where water comes through the machines. It stops the dust.

RICK: But there's still so much dust in the air from the blasting and stuff—there's always dust. We have a lot of problems with those insurance people and the government regulations, but they've saved a lot of lives.

Like with the skips. They used to call them coffins. They were just a cable with a metal box, twenty-four inches wide and twenty-four inches deep. You could just barely fit in there.

ROY: They're like a little elevator. You could ride a skip 200 feet up the raise. Follow the vein up from level to level.

RICK: It leaned, because the vein would be on a slant. The veins usually laid at about eighty-five degrees. You always had to lean back a little bit because all that guided it, as it went up, was just a two-by-twelve and then a two-by-four runner on both sides.

One of our friends we grew up with, Gerald, he was working over at the Idarado. He was riding up with some parts or something. One guy used to ride on top and manage the cable. The other guy would be inside. They were up so far, and the cable broke. Just dropped them.

ROY: When Gerald came down and hit, the impact took his leg. He lost his leg. He was just lucky to survive it.

RICK: The same thing happened with Ervin. At that time they'd updated the skip. OSHA [Occupational Safety and Health Administration] had come in and told the mines they had to start putting guides on the sides of the skips. They had "dogs," they called them, that were spring-loaded and were supposed to lock in to keep the skip from dropping if the cable broke.

But they hadn't been cleaned. When the cable broke, the dogs cracked too. Ervin only dropped about thirty feet, but you stop in a jiffy. He wound up with a compound fracture in one leg.

There were lots of accidents everywhere in the mine.

ROY: I was really scared one time. I came out of the raise, the manway. You have an air hose that supplies the machine, and mine had a bubble on it about to break. Well, I sat there and I fixed that.

I walked over to go pick up my diggers and my coat. Turned back around and was walking toward my machine, which was about twenty feet away, and a slab about the size of a picnic table, came down exactly where I had been sitting, doing all that work. I think somebody was looking out for me that day.

RICK: It just wasn't your time. You never knew what could happen. The conditions outside were miserable too. It snowed all the time, and we drove our own vehicles to the mine.

ROY: We'd all chip in and buy an old car together. Drove that back and forth to the mine.

RICK: We came out one night, and it was cold. There was one spell there in Silverton where it would be fifty degrees below zero for three or four days. We would come out from underground, wet from mining and stuff. We had a dry room, but the water was usually bad.

ROY: The dry room is where you usually change your clothes, when you come from underground.

RICK: When we came out, our old car wouldn't start. We were on night shift. It was so cold and miserable. There was a hill coming out of the portal to the mine, where we had parked. We got out to push the car; but it was so cold, that the brakes and everything had froze up. We couldn't even push it down the hill to get it started. That was cold!

ROY: We were on the mine rescue teams. If something happened, they'd call the mine rescue, and you'd go and get the person out. We'd train to handle situations.

We would compete against other rescue teams from around the country. Mock mine disasters. Getting ready for situations, if they ever happened. Know how to handle it, know how to rescue guys, know how to do first aid on them.

RICK: A lot of it was fire training, for if a fire got going underground. You had to be prepared.

ROY: Like when those tourists built a campfire in that mine tunnel near Eureka. They thought it was neat because the smoke would actually go into the mine, and so they didn't have to breathe the smoke.

RICK: They thought that was pretty nice. They didn't know there were people working back in there.

ROY: About sixty-to-seventy guys were working in the mine.

RICK: Smoke is deadly underground, because you don't have that much air anyway. By the time you usually see smoke, the carbon monoxide around will kill you. The first sign of smelling smoke or seeing smoke, you got to put your respirator on. They call them "self rescuers."

ROY: But the greatest fear is bad ground—slabs falling while you're drilling. They never did make anybody work someplace they didn't want to go.

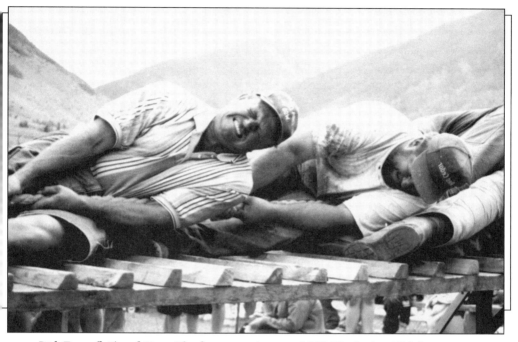

Rick Ernst (left) and Terry Rhoades in tug-of-war at 1999 Hardrockers Holidays.

RICK: Most of the bosses were miners first and just moved into boss positions, so they knew the dangers as well as we did.

ROY: Sometimes you did get guys who just came out of school—"greenhorns" that thought they knew it all. You just said, "Whatever," and then you did it the right way.

RICK: As far as being book smart, they were. As far as having experience and stuff, it was totally different; and most of them, they learned. Any of them that were worth a damn.

ROY: To others we were just dumb miners. There was a geologist that came into our heading one time. He wanted to do this and that. I said, "I tell you what, Bud. If you don't get the hell out of here, I'm gonna take you, and I'm gonna put you in this muck car, and I'm gonna dump you down the ore pass." He turned around and walked out and complained to my partner. My partner said, "You know what, whatever he told you, I agree with. Why don't you get the hell out of here." Never heard anything more about it.

RICK: They had bosses in the early years who were great, but the later years they brought a few guys in that had been to the School of Mines for mining management and all that. To go through school, you learn a lot of that, but you don't know what it really takes until you get in there and do it—the mining, the blasting. There's a lot more to it than what you read out of a book. If you have never been underground, it's a different experience.

ROY: I still remember my first time underground. It was neat. We'd go in two miles on the mantrip. It's like a little train that's thirty-six inches deep. Just like a narrow gauge. You'd go in two miles, and then you'd take the skips up to your different working levels. You'd turn your light out, and you couldn't see your hand in front of you.

RICK: That's what really got to me when I was underground. When you had your light on, you could see certain areas where your light was shining; but when you were by yourself, and your light went out, or you turned it out, it's totally black. There's no nothing. You can't believe how dark it is. I mean you don't even know which direction you're walking, which way you're going.

RON: In the main areas they'd string lights, around the shops and the dog house, where you ate and stuff.

RICK: People were always trying to scare you or mess with your mind by shutting off the lights. There were guys in there, we'd capitalize on their fears. One guy, Bobby, he hated sardines. I mean he'd just go to throwing up, even just talking about sardines.

Almost everybody had a can of sardines at lunch. One day, before Bobby had gotten to the dog house, a guy opened up Bobby's lunch pail, then opened his sandwich up, put a sardine in the middle of his sandwich, and put it back in the baggy, and we all just sat there waiting. Bobby took a bite, and he chewed, and he chewed, then—ahhh!—he threw up.

ROY: One time I took that kids' stuff they call green slime, which used to have eyeballs in it. I opened Bobby's lunch pail and poured that green slime across the top. Put those two eyeballs looking straight up. He came in for lunch. Lunch was a special part of the day. That's when you bullshit. You look forward to that and not working. He came in and opened that and screamed, "Oh, my God!"

RICK: You didn't want to show any weaknesses. Guys would take advantage of you. You made sure you never fell asleep. Guys had ammonia inhalants that you get out of first-aid kits, used for reviving people when they fainted or went unconscious.

Some guys would pack those on the mantrip going out, on night shift. You always had to wait for the mantrip to fill up. Everybody had to be there before you could leave underground, and some of the guys would sit out there and sleep. Everybody would have put in a full shift, and they'd be sitting there and fall asleep. There would always be somebody with some ammonia inhalants, who liked to play a joke. Talk about bringing a guy out of a sleep. Oh my God! Come up a fighting, boy! That was always a good one.

One that they'd do all the time to the new hands and the nippers and stuff, is they'd send them up to the shop to get stuff that didn't make any sense, like left-handed monkey wrenches.

ROY: You'd say, "Go to the shop and get me a left-handed monkey wrench. This one here's a right-handed one. It won't work!" Of course with wrenches, there is no left or right, doesn't make any difference, but they'd go up there looking for a left-handed monkey wrench, being new and gullible and all.

RICK: Somebody would tell them something, and they'd just go do it.

ROY: They didn't know how to dress. They'd have their diggers on all crazy. Their light in a tangled up mess. We had fun with them. Mining was dangerous, because you're drilling and blasting, but it was also good times.

RICK: I loved every minute of it. I have no regrets about it. We worked hard, but you had a lot of good friends. And even though some of our friends got killed there, it was a good job.

It was just like a family. There were never really any problems with the kids. Everybody knew what everybody was doing, so if your kid messed up, it got straightened out.

RICK: When Lake Emma caved in and flooded the mine, that was the beginning of the end, right there. The lake was over the mine. I guess they'd had us mine up too close to it. It was real lucky it waited until Sunday to cave in. It could have happened any other day of the week. Nobody was underground Sunday. The usual shift was about 125 people. It would have killed everybody.

ROY: It could have been one of the worst mining disasters ever. The guys that did survive the initial flood, they wouldn't have been able to get out.

RICK: The mine shut down because of the flooding. It was around 1978. A lot of people moved away. It took months and months of cleanup to get in. Tons of debris came out of the bottom of that lake.

ROY: During the cleanup, we worked for seven bucks an hour—we were used to making about twenty dollars an hour.

RICK: I was lucky, because my parents and brother were down in Nucla. I got to go down there and mine uranium. I'm still mining.

Eli Romero

Durango, Colorado

My father worked for the Standard Metals Mine. He didn't want me to work there. He wanted me to go to college. But I had mining in my blood, and I saw the money he made. I started there in 1973.

In those days, everybody depended on the mine, one way or another, and a lot of it was family. If you weren't a Romero, you were related to one of us; or your cousin was married to a guy who went to high school with us. Everybody pretty much grew up together. When an ambulance went out to the mine, you wondered who'd been killed or hurt. If someone got hurt in the mine, we'd put money into a pool. We took care of each other in that way. The only injury I ever got in the mine was getting the tip of my finger cut off. I had my finger on the chute when the rock came down.

Everybody loved mining; but, at the same time, they'd be out of there if they had something better going. It was the money. Miners got paid by how much rock they broke. As trammers, we got paid by how many ore cars we hauled out and dumped down. A lot of people never brought lunches. They just worked right through their shift.

You would go in early, like at 7:00 in the morning, when it's dark. You'd go home at 4:30 in the afternoon, when it's almost dark again; and, in between, you're underground where it's always dark.

Sometimes in winter the power would get knocked out underground, and you'd have to climb down the ladders, all the way down the main line. You had head lamps, but a thousand feet is a lot of ladders. Some guys actually tried to do it with gold in their pouches. Crazy, but I did that at least once. I said, "If I'm going to walk out of here, I'm going to walk out with $600 bucks."

The mine broke only twice a day, between shifts, to blast out the gasses. It gets really nasty underground. I ran a lot of marathons when I was working in the mines. I'd be running and coughing out black gunk.

The real depressing part of working in the mines was that you start out as just a flunky, work your way up to mining, and then after mining for so many years, you get conned up—it's called "miners' con," which is like black lung. You lose your strength. You can't move that well or handle the heavy drills. Basically you can't handle being a miner any longer. So you take lesser jobs, until you're back down to a flunky again.

There was this old man named Billy Hunt. He mined for a while, then his lungs got bad, and he was back down to blasting with dynamite. But he was so slow that after he lit the fuses, he'd just be getting down the ladder when the blast would blow him off it. This guy was about seventy-five years old. He'd kind of come out of it and say, "Okay, tomorrow's another day."

Billy had a dog, and they would go downtown on Saturday nights. He would carry a tray with him for his dog. They'd each drink a six-pack and then stagger home together. It was a funny sight, but one day the dog made it home, and Billy didn't. Billy couldn't make it over a snowbank, so he froze to death.

My father died of miners' con. When you're drilling with water, most of the dust is knocked down; but there's such a thing as drilling dry. It's against the law now, but a lot of miners did it years ago. My father did it. It's where you don't hook up the water, because there's problems with the water, and you want to keep working.

My father got a cold one day. He was coughing real bad, and he went to the doctor. The doctor said, "That's it, you're done. You can't go back to the mine." My father was only forty-seven years old. He said, "I got to go back. This is all I do." But they wouldn't let him.

I was on the grievance committee for a while. Me and my brother-in-law and some other boys wanted to strike for better health and life insurance. We used to meet at Romero's, the restaurant my parents started before my father died.

The grievance committee would sit there every day, drinking margaritas and getting looped, and arguing about how come these people were making more money, and nobody was making money, and back and forth. After so many margaritas, some of us would start to fistfight. Well we decided Romero's wasn't the place to have the meeting, so we moved it down to the American Legion, and there they drank whiskey. So that got crazy too.

The mine flooded around 1978, when Lake Emma caved in. Thank God it happened on a Sunday and nobody was underground, but suddenly everybody was out of work. Some of the miners left town, went to other camps. Others of us stuck around and when it came to cheap labor to clean out the mine, they had us over a barrel.

We went back in there to do the mucking. After the mine caved in, the whole thing was filled with mud. The main line, everything. We'd put a chalk mark on the wall where you started, so you could figure out how far you got. Well, sometimes you wouldn't even see the chalk mark because it would keep oozing out and pushing toward you. It was dangerous and pretty nasty.

The town went from 1500 to 500 people. They combined the classes in school. I decided to move my kids down to Durango to get them a better education.

There's still gold up there. It just hasn't been discovered, and it probably never will be, because the environmentalists won't let it happen. There are so many permits. Also, when the price of gold fluctuated, and the mines were locked into a price they didn't like, mines like Idarado pulled out. They ended up just leaving all the gold in there and shutting it down.

The thing I don't understand about environmentalists, and about mining, and about the whole thing, is that they're trying to purify the water and bring back the fish. But there were never fish up there, in the sections where there's too much mineralization, and the environmentalists think they're going to change this.

Mining is a lost art in Colorado. There's still mining in other places, but it's really strict here. I don't know why. The mountains, I guess.

Terry Rhoades

Silverton, Colorado
(Billy Rhoades's son)

When my dad was leasing up at the Argentine Mine, he worked weekends. I could go up there and watch them do everything. So when I turned eighteen, I was ready to go mining. He was at Standard Metals then. Dad didn't want me to go into mining; but when I did, he wanted to make sure I was safe, so he stuck by me.

A lot of the old guys wouldn't mind taking young guys with them, to teach them how. If you got a great big kid, like Rick Ernst, who was willing to do anything, they'd think, "Well, if I teach this guy something, we're going to make some money." And they did.

You treated the old guys with respect. If you shined your light in some of the guys' eyes, they'd knock your light out. You learn. I was a lucky person to get to go mining when I did. Some miners had reputations that preceded them. Rod MacLennan was one. They said if you ever go into a drift with him, you'd better be able to drill, or he'd be yelling at you. If you were lucky enough to get a go with an old-timer, you went in there and busted your butt.

My dad took me on, mining, because he wanted to look after me; and he was pretty cautious with me, but we made good money. We both worked hard. I mined with him for probably about a year.

Everybody around here looked up to my dad. People liked him as a shift boss, because he didn't ask them to do anything that he wouldn't do. He was small, but he could do anything anybody else could underground.

He'd just have to use his machine differently than a big guy. Stick the drill's leg out farther and stuff.

In his younger days he was a fighter. He was a Golden Glove in the Navy, then he was a bar brawler. He fought all the time before he was married. I guess we straightened him out.

One time we were drilling on a raise, and he told me to grab a single jack and hit the steel, while he was twisting it. I hit it a couple of times. He said, "I told you to hit the son of a bitch!" I lifted the single jack back, and I swung hard and hit him on the hand. He wasn't too happy at all then.

My dad got hurt a lot of times. I would be out hunting with his mining partners, and I heard stories. He got buried once, all the way up to his chest. He had a lot of accidents. The last time that he was hurt, he was on the main level. He was helping somebody hook up the cars. He got his leg caught in there, and it broke his leg in about ten places.

It was understandable that he was concerned for me. He rode me and my partners hard. He kept saying, "Just make sure you bar down. Make sure you're safe." That was his main thing with me, being safe. One time when I was on the raise climber, at about 600 feet up, a slab came down and hit me. At that time you didn't wear safety laniards or anything like that, where you hooked on. If I wouldn't have hit those steel rods in the put-ins, I would have gone all the way down.

After that, and breaking my hand once, my dad was real scared for me. He wanted me to go somewhere else for a while, so I could see if I wanted to do another kind of work. So I quit and went to work on an oil rig down in Texas. I had to work seven days a week, twelve hours a day. I had made a lot more money mining, working eight hours a day, so I came back, and my father was happy about that.

Most of the people I know who have been killed underground were hit by slabs. The pockets in the scrams would get hung up—all the big boulders hanging up in there. You had to crawl up on ladders, put dynamite around to blast them down, so they didn't fall on you by accident. That can be a little scary. One time I was running out of there with boulders coming behind me. I hit my head on the rail bolted at the brow of a pocket, and flipped. I quickly crawled out of the way of the rocks coming down, but I wound up getting a bunch of stitches on my head.

With Johnny Castle, I guess he just got hit by a really big slab. They were going to bolt up that slab. That's where you drill a hole through it and put these pins in it that hold it up there. He went to bolt it up, but it broke behind him, fell and killed him.

My good friend, Steve Davidovich, was barring down in a chute with a drill steel. You're not supposed to use steel, because steel doesn't bend. You're supposed to use an aluminum bar. Anyway, he hit that rock with the steel, and that rock was probably about 600 pounds. It caught the steel, and the other end of the steel caught him underneath the neck. Broke his neck.

I had to go tramming right after Steve was killed, because they couldn't get anybody else to go in there. That's one time when a guy starts thinking a little bit. Hell, two weeks later, I had the same thing happen to me with a bar, but it caught me under my arm. It took me up against the back; but a bar will bend, so I was okay.

I hear things in the mine. Say you're tramming, and your partner drives the train out. You're back there alone in the dark. You hear a lot of things, or you look down the drift and think you see a light, but it's usually the reflection of your light going off the water down there. It was easy to scare people in those conditions.

One time I crawled up the ladder. It took me about half an hour. Hippie John was slushing in a scram. I had this idea that I'd scare him. When I did, I thought he was going to kill himself. He jumped and started screaming.

We had this one guy who would hide in the powder box, when he knew the guys usually came to get their dynamite powder. He laid in the powder box until they came. When Bobby Gallegos opened up the powder box, he almost passed out.

Terry Rhoades at the Hardrockers Holidays.

When some of the guys couldn't mine anymore for health reasons, the bosses would put them on mine repair, which included a lot of timbering, so they still had to work hard. Usually they were conned up. "Miners' con" is what they called it. Silicosis is what it is. They used to make you take x-rays for your physical, to see how many spots you had on your lungs. With the older guys, it had a lot to do with if you mucked or drilled dry, like my dad used to do sometimes. He eventually died of lung cancer.

It was also just plain dirty in there. Sometimes the diesel was so thick you couldn't even see three feet ahead of you. At one time MSHA (Miners Safety and Health Administration) made it mandatory that you wore a respirator, but nobody did. It was too much of a hassle. Then they came at us with the safety goggles. When you're drilling up above you, all the water and cuttings come down in your face. It takes about half a minute before your glasses are all covered with muck to where you can't see. It was actually more unsafe to wear them. It was a noble thing to say they're good for you, but it was impractical.

We all had a close call when Lake Emma flooded and caved in the mine. The engineers had actually tested it under the lake for depth. They figured there was a fault going up there, because there was real bad water coming through. They didn't realize it was half full of mud.

The guys that were driving that stope—I think it was Fred and Harry Castle—they refused to go to work that Friday night because they said that water was pouring out of there. That Sunday it came through. Luckily nobody was underground. There were places, like where I had been working, where the mud wouldn't have got to you, but there was no way you would have been able to make your way out of there. You would have starved to death.

It took over a year to muck out that flood. One day I was with my dad, mucking out some pockets on the main level, when all the mud and muck broke loose. I started running, and it hit me in the back. Took me down the drift about 300 yards. I lost my light. It buried me. I finally got out, and I went back looking for my dad. The only thing about the underground I know for sure is that you look after your partner at all times. He had jumped on the mucker, and he was okay. The mucker weighs about eight tons. The slide had taken it off the track. Knocked all the cars off.

So we cleaned up the flood and went back to mining a year later. We were glad to have the work. But then the mines around here started shutting down, mostly because the price of metals dropped. I miss it.

Mark Parker

Norwood, Colorado

My old man worked for thirty some years at the Climax Mine near Leadville. He had some strings to pull, so I kind of fell right into it. I remember the first time I went in, I was very conscious of being underground. I thought I couldn't breathe. I had a hyperventilating thing going on. But there was thirty or forty other guys with me, and they were all fine. They were smoking cigarettes, joking, working, and everything else. I was the only one who was dying. I kept looking up, because I thought the whole world was going to fall in on me. I went over, and I picked on the rock. And I knocked on it with my knuckles and thought, "Well that's pretty hard. I guess maybe it won't fall in, today at least." I don't even think my nervousness lasted more than half a shift. It never bothered me again.

But I didn't feel Climax was for me. When I was working there, they had 2,500 people underground, and that's too many people to keep track of. I never had a full-time partner. When you're mining, your life was in many other people's hands, and some of the people who showed up there, I didn't like the feeling of them.

There were quite a few accidents. We had good miners mixed with people who didn't care if they were flipping burgers at Burger King or mining. It didn't matter to them as long as they got paid at the end of the week. I probably had a few more close calls than I even knew about.

I stayed there four months, maybe. I didn't want to set myself up permanently, so I didn't get a place to live. I would camp out there all week and then come over here, around Montrose County, trying to rustle up a mine job in the area. I couldn't understand why they wouldn't hire me. Looking back now, I can see it. I was eighteen years old, fresh out of school. I didn't know anything, and I wanted to be a miner? Come on— they needed grown men who knew what they were doing. They didn't need school kids; but they finally hired me, because I kept bothering them. I started at the Sunday Mine, mining uranium, in 1977.

Miners know miners. Usually your reputation gets to the job before you do. I have yet to find a mine that I've hired on with, where that hasn't happened. They know all about you. They know what you've done. They know who you run with. If I would have been a mining boss, and I was doing the hiring, I'd listen to my hands—my workers—real hard. Miners know good hands. We know hands that are okay; and we know hands that, well, they're okay guys, but we really wouldn't want to be on a contract with them.

Most places today, they hire out of the job service. You've got to go through some kind of unemployment set-up, and they wouldn't know a miner if he jumped out and bit them. You write down on a paper what you're qualified to do, and they read it. They tell the employer. That's crap. I'd hate to be a mining boss now and have to hire people that way. That's not how you get good miners.

The Sunday Mine lasted until 1983 before our first layoff. That was during a major shutdown on all the uranium industry in North America. It seemed like we all went belly-up at about the same time.

I went to work for Idarado doing the reclamation. That's what I'm still doing. The Idarado had been a hardrock mine, until it shut down in 1978. They can't just blockade off the portals, because of the water. We have to keep it under control for reclamation and environmental reasons.

It's sad that the mine is so empty, because I believe that mill up there is the biggest of the old mills still standing on the Western Slope of Colorado. It's sad to go in there and think that at one time, my God, a big part of Montrose, Ouray, and Telluride all depended on that one mine. That was the livelihood. Probably one of the biggest taxpayers in the state, and now we're down to seven guys.

We don't have a day-in, day-out partner like we used to have. We're a small crew, and we go wherever we're needed. I think there's what we call "tommyknockers" under there—ghosts. I hear noises, there's no doubt. And if you start thinking about tommyknockers, you'll start thinking about all the other spooky things under there. Pretty soon you can scare yourself. You let your mind wander a little bit too far, and it'll make your ears stand up once in a while.

A few times down there, I thought somebody was looking at me. I thought I heard something. You think, "Oh, you're foolish," but who's there to tell you you're foolish? You're by yourself, or you wouldn't be scared in the first place. If you had a partner with you, it wouldn't have bothered you a bit. Lesson to be learned, keep your partner with you.

What was a concern was the factor of the mine collapsing. You had to constantly worry about that, keep it in your mind, and control it. Stagnant mines become unstable. They deteriorate very quickly. Those timbered chutes and stuff have been sitting there since 1978. Things have a tendency to rust, rot, and just get old and tired, including old miners. Things start giving way. There are cave-ins, places in the mines where we just can't go anymore.

They think this reclamation won't go on forever, but who knows. How do you stop water from coming out of the San Juan Mountains? You can stop it one place, but then it has a tendency to go in another direction and come out somewhere else.

I'm no specialist on water, but I do know it goes downhill. And if you back it up far enough, it usually takes its own path after awhile. It would be a perfect plan, if they could stop everything, and all the water quality was great, and there was no stain on any rocks or anything, but I don't see how you can do it. We have to have someone here maintaining all the time.

People want to change it all back to the natural way it was 100 years ago; but I think that 100 years ago, if there was some guy down the river taking samples with his coffee cup, he might have found out that there was as much zinc and lead in the water as there is now. The only thing is, we didn't have those people back then. The people who were up there came to work. They were too damned busy to be down there taking water quality samples.

So, all of a sudden now, somebody reaches down in the creek and takes a dip of water and tests it. They say, "Jesus Christ, this water is terrible. We've got to do something about it." If you asked Chief Ouray how the water was back then—he was living over there at Red Mountain—I bet he would have told you it ran red every fall when the big rains come. The creek is red, no fish.

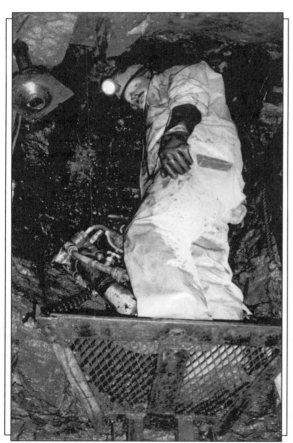

Mark Parker working in a raise.

When all our mines in the San Juans were shutting down, Nevada was just getting a good start. They found out they had massive microscopic gold deposits. Acres of it. All of those big, open-pit mines came in. In the pit mines, they're running 250-ton trucks, giant shovels. Underground guys can't compete with that for productivity. The underground hardrock miners were the last of an era.

Unless you were lucky, it used to be that mining was the only way a person with no extra schooling could make a fair living. Employers paid you for what you actually did. Nowadays there are hardly any jobs like that. It doesn't matter if you're the guy who can't or doesn't do anything. Some still get the same amount per hour as the guy who's getting it done. With mining, the more footage, tons, or cubes you got, the more money you made. You actually had a chance to, I don't know if I would call it "improve yourself," but you had a chance to up your financial status by hard work.

There was a lot of pride involved in doing a good job. By God, if your name was on the top of the contract board, that was something to brag about. We'd go to the saloon that night, have a beer or two and say, "Yeah, we topped the board this time." It's a prideful thing. We all tried to do the best we could and be the best we could.

There were a lot of people who shunned us, but we didn't care. We thought we were the kings. We were the ones who mined the metal that people sat their butts in to go down the highway at ninety miles an hour. We were the ones that mined the gold that hung around their necks, but they weren't smart enough to ever figure that out. Where did they think gold comes from? Somebody went and got it out of the ground.

If you ever actually picked up a piece of rock and saw gold, it is pretty awesome. Something about it just grabs you. There are not a whole lot of people in the world who have ever got to go in and pick up a piece of rock, turn it over, and see gold, the way it was meant to be.

Aaron "Bootleg" Calhoon

Ouray, Colorado
David Calhoon's grandson

My grandpa, Dave Calhoon, he was a miner, and that's what I wanted to be. I was twelve when I first went into a mine. You're supposed to be eighteen. I loved it. You're breaking rock, and you're in new ground every day. You never know what you're going to find.

I got the nickname "Bootleg" because when you drill a round out—if you shoot it, and it doesn't break the full depth of the holes—that's called a bootleg. All the holes will just shotgun rifle out. You'll have a bunch of big holes and there won't be a muck pile there. I did that the first couple times I started mining, so they hung that name on me.

My dad never was interested in mining. I took him underground in the Grizzly Bear in Ouray when I was in high school. That was the first time he was underground.

I'm going to school for my mining degree. It's a four-year program. We have to take chemical, civil, and mechanical courses, plus calculus, physics, and chemistry. Then we go into specific departments. People tell me it's a dying business—don't go into it. All my teachers in high school were always trying to talk me into going to school to learn something else, but my first year mining, I probably made more money than they ever will.

After I graduate I'll go back to mining, but I don't have to. I can go off into other branches of engineering. I have options, in case I get to be forty years old and decide I hate mining.

There'll never be anybody like the old miners again: Billy Young and those guys like Pete Klein and George Munzing. In Nevada right now, you don't have to be a really good miner to mine, but back in their days, when they were mining up here in the San Juans, you had to be good.

My partner, George Munzing, says that when he started it was the older guys, like Bill Young and Walt Orvis, who showed them what to do. There were no new mines opening up. Guys getting out of high school, they still went to work in the mines, but they weren't on the big contracts. It was just the old guys mining, guys in their thirties and forties. When I started, the older guys just kind of took me under their wing. I didn't have to go through what they went through.

Everybody that works underground, they're pretty unique. I've never seen the same sense of humor anywhere else. How we joke underground, if I acted that way with the people I go to school with, they'd think I was nuts. Swatting, or poking, or marking **x**'s on people. It's just different.

Mining's easier now, because of technology. You're not doing a lot of timbering anymore. You don't have to worry about powder explosives as much. They're more stable now. They still make a nitro-based powder; but, unlike before, it can fall down by mistake and not go off.

I worry about the diesel smoke. I hate rubber-tired mining. That's all diesel equipment. I like track a lot better, because you can use electric motors. After a time that diesel smoke is bad for you. I guess I'm too young to worry about the arthritis and the other things people get from working in the cold and wet.

They still have a lot of accidents. I have an MSHA [Miners Safety and Health Administration] report that shows the fatalities through September of each year. In 1997, they had fifty. That's a lot, because there's not that many people mining anymore. I didn't think that was too good.

I don't know if it's luck, skill, or just fate, what happens to people. There's been a lot of good guys who I would never have worried about, who got hurt or killed. There's only a few times that I think it was just lack of knowledge.

What happened to Joe Mattivi shouldn't have happened. But it did. I was working with him over at the Grizzly Bear. He was out on the surface. He was a mine mechanic. They made him start dumping their muck right beside the portal. He was going in there with this little tram motor, mucking it, hauling it, and dumping it. He must have had a hydraulic leak. He put a wrench on the hose. He probably thought it was loose, but it actually had a hairline crack. It popped off and busted all the hydraulics. The tram bucket fell on him, and killed him.

I never thought anyone would ever be hurt at the Grizzly Bear, because they weren't on contract, nobody was in a hurry. They were getting paid by the hour. I was the one who found him.

Joe Mattivi was always really careful. If anything would have happened, I would have thought it would be underground. He barely outran a cave-in that killed one of his partners over in Silverton. He didn't like going underground anymore, but Joe wound up getting killed anyway.

I've heard about ghosts or spirits underground. People talk about it. When you get to the intersection of the Camp Bird Mine, it splits. It goes east and west. The west Camp Bird is the old part of the mine. People see ghosts in there. Nobody likes going in there.

Part of me wishes I was born earlier, so I could have been there for the boom. There will be a few mines around here still, but I'm sure Silverton will be a tourist town, like Telluride, in another twenty years. I'd rather go down to South America where they're really mining. I'd love that.

The toilet at the Old One Hundred Mine.

GLOSSARY

This glossary compiles information gleaned from various sources. David Calhoon and Al Maes assisted a great deal with vital information. It should be noted that mining terminology varies greatly between different mines and mining areas.

Amalgam: A mixture of mercury with a metal, like gold or silver.

Amalgamation: The act of mixing mercury with a metal, like gold or silver.

Bar: A scaling bar; a long rod-like implement with a chisel at the end.

Barring Down: Using a bar to pry off potentially hazardous overhanging slabs or boulders.

Base Metals: All metals (excluding precious metals like gold, silver, and platinum); such as, lead, zinc, and copper.

Belt Line: A fast, power driven, horizontal belt that is about six feet wide, is raised off the ground, and runs on rollers. A belt line is used for moving coal or rock out of a mine. It will sometimes get out of balance and ride up on one side or the other, spilling the material. In such instances, the person working the belt line has to retrain the belt so it rides in the middle of the rollers again, and then shovel the coal and rock back onto it.

Bit: Drill bits were originally part of a the drill steel, but later were made to screw on the steel, so they could be easily replaced.

Bull Hose: A large hose, about two inches in diameter, connecting high-powered compressed air to mining machinery, like a mucking machine.

Carbide Lights: A miner's light or lamp with a reflector lens that uses carbide, a compound of carbon element, to fuel the flame device. It is usually mounted on the front of a miner's hard hat for the purpose of illuminating the work area. Carbide lights were used prior to battery powered electric lights.

Cat: A bulldozer, usually but not always Caterpillar brand.

Chute: An incline or opening, from one level to another level, for dropping ore into cars for transport to a different area.

Collaring Holes: Using a drill to mark the pattern and location of the holes that will be drilled, with various lengths of steel and drill bit sizes, and then loaded with explosives.

Come-alongs: A hand winch; a hand powered hoisting machine with a drum around which a cable is wound. The end of the cable attaches to the load being lifted.

Commissary: A company store.

Compressor: A machine used to compress air for powering machinery.

Contract Mining: When the mining company pays a crew for the amount of work accomplished in a given amount of time (such as, so many feet of drift), as opposed to a predetermined hourly wage.

Coupling: A device used to link two things together.

Crosscut: A level or tunnel driven at right angles to the main tunnel or vein.

Diggers: A) Work clothes similar to overalls. B) A machine used for excavating.

Dipper: A) A loader bucket. B) The lip on the front end of a loader bucket.

Dog House: A small excavated area in the mine, enclosed with boards, where miners go to eat lunch and dry wet clothes.

Drift: A horizontal, or near horizontal tunnel underground, which follows an ore vein.

Drifting or Driving Drifts: Excavating a drift, including drilling, blasting, and removing broken rock.

Drilling a Round: Setting up and drilling enough holes to advance the drift. The holes are then loaded with explosives and blasted.

Drilling Dry: Drilling without infused water, so that dust particles are more airborne than usual. Dry drills were nicknamed "widow makers."

Driving Tunnel: Excavating a horizontal passage underground, coming from the surface.

Face: The end of a tunnel, drift, or excavation, where the work is done.

Floatation: A process employed in the mills using certain chemicals (usually detergents), mixed with water which, when frothed with paddles to form bubbles, tend to attract finely crushed ores. The bubbles can then be scraped off the surface and collected. Different chemicals attract different ores.

Foot Wall: The lower wall or side of an ore vein.

Graveyard Shift: Shift times vary in mines. The graveyard shift is approximately midnight until 7:30 in the morning.

Guide Shoes: Guiding devices on either side of a skip. They run along four-by-fours on the inside of a shaft to keep the skip upright and in the middle of the shaft.

Hanging Wall: The upper wall or side of an ore vein.

Hardrock: Rock containing hard metals like gold, silver, lead, and zinc; not soft minerals like coal or uranium.

Head Frame: A wooded or metal structure built over shafts and enclosing a hoist.

Heading: Same as Face.

High-grade: Ore of superior quality, rich in gold or silver.

High-grading: Removing or stealing high-grade from a mine.

Hoist: A lift; a power driven drum with steel cable attached, used primarily for raising or lowering miners and mining materials to various levels.

Hoist Man: Hoist operator.

Hoist Room: Location of hoist equipment and hoist man; may be on the surface or underground.

Jack-leg Drill: A drill with a metal "leg" that jack-knives out to partially support and steady the drill. When drilling, the miner uses the leg to apply pressure on the rotating steel.

Lode: An irregular vein without well-defined walls.

Luddite: One who opposes technical or technological change (after Ned Ludd, an English laborer who was supposed to have destroyed weaving machinery around 1779).

Mancage: Also known as a cage. A contraption similar to an elevator, used for hoisting or lowering ore cars, men, and material.

Mantrip: Rail-mounted cars constructed with benches to transport miners in and out of a mine at the beginning and end of each shift.

Manway: Usually a vertical excavation about the size of an elevator shaft, with ladders for climbing to different working places (stopes), or to different levels.

Mine: A large excavation made in the earth, from which to extract metallic ores, coal, precious stones, or certain other minerals. May be open pit, or underground.

Miner: A person whose work is excavating coal, ore, etc. in a mine.

Motor: The locomotive that pulls the mantrip or ore cars.

Motor Man: The engineer who operates the motor.

Muck: Earth, rocks, or clay excavated in a mine.

Muck Bucket: Usually a round barrel, used for transporting muck up and down a shaft.

Muck Car: A rail mounted car used primarily for holding and hauling rock and ore.

Muck Train: Two or more muck cars coupled together to move rock and ore.

Mucker: A person hired to operate a mucking machine, for removing muck; also used as a nickname for a mucking machine.

Mucking: Operating a mucking machine.

Mucking Machine: A power-driven machine, which runs on rails, and has a scooping device in the front for picking up and dumping muck into containers for removal.

Nipper: An entry-level job in a mine. Nippers usually work in teams, moving supplies and materials to different working areas.

Ore: A mineral or an aggregate of minerals from which a valuable constituent, especially a metal, can be profitably extracted.

Ore Pass: Similar to a raise or a shaft, but without timber. It is used to drop ore to lower levels.

Ore Pockets: Storage area for ore or waste.

Pillars: A block of ground left in place to support the hanging wall in a stope.

Portal: The opening into a mine.

Powder: Explosives used in a mine.

Pulling Pillars: Blasting pillars after excavation of ore is complete, and safety is no longer an issue, in order to get the remaining ore out of a stope.

Prospecting: Hunting for mineral deposits. Prospectors usually sold their finds, or mining claims, to mining companies, which had the staff and machinery to excavate it.

Put-ins: Small holes in the deck where you put spare steel rods for drilling.

Raise: A vertical or near vertical excavation, about the size of an elevator shaft, driven from the bottom up. Similar to a manway.

Raise Climber Machine: A cage-like contraption that runs on a rail on the hanging wall of an ore pass. Used for transporting miners to their working areas, the raise climber has a trapdoor above, so the miners can get on top of it and work. The rail also conducts power, air and ventilation to the working area.

Reagents: A substance used in a chemical reaction to detect, measure, examine, or produce other substances.

Scrams: A small tunnel where they run the slusher buckets, to muck up the ore pockets.

Shaft: A vertical or incline excavation for prospecting or working mines. Commonly the size of an elevator shaft, it is sunk from the top down.

Shift Boss: A front-line supervisor.

Shifter: Same as Shift Boss.

Single Jack: A four pound hammer.

Skip: A square bucket used to raise or lower rock, supplies, or men. It travels on wooden guides in the shaft. The term "skip" is often used interchangeably with mancage.

Slabbing Off: Drilling holes on the sides of a drift, and blasting them, in order to widen the drift.

Slush: Soft mud; mire. In a mine, slush refers to broken rock and ore.

Slusher Bucket: A cylindrical vessel or scoop that transfers rock or ore horizontally or on an incline. One commonly used method employs a double drum hoist with a cable for pulling the scoop in one direction and then back in the other direction.

Slushing: The process of moving slush from one location to another.

Steel: A steel rod that goes in any drill, like a stoper, for drilling.

Stope: An excavation from which ore has been extracted from a vein.

Stoper: A rock drill intended for drilling vertical holes.

Stoping: Breaking ore from a stope or section of ground in a mine, between or above levels.

Swing Shift: Shift times vary in mines. Swing shift is approximately 4:00 in the afternoon until 11:30 at night.

Tailings Pond: The run-off of chemicals and inert matter from the mill, which is contained by a dam to form a pool, from which the water eventually evaporates.

Timbering: Providing ground support with either wood or steel.

Tram Room: A room on the loading dock of an aerial tram, or tramway.

Tramcar: A car, similar to a large bucket, that runs on rails, for transporting broken rock.

Trammers: Men hired to move broken rock.

Tramming: Moving broken rock with a tramcar.

Tramway: An aerial tram. Tramcars that hang from a cable, for transporting men, materials, and ore across a canyon; for instance between a mine and a mill.

Trolley: A small truck or car operating on a track in a mine.

Trolley Pole: An apparatus for collecting electrical current from an overhead wire and transmitting it to the motor of a mine locomotive through contact of a wheel at the end of a pole.

Vein: Any mineralized zone.